W9-CKF-439

The Concept of Nature

The Concept of Nature

ALFRED NORTH WHITEHEAD

NEW YORK

The Concept of Nature
Cover © 2007 Cosimo, Inc.

For information, address:

Cosimo, P.O. Box 416
Old Chelsea Station
New York, NY 10113-0416

or visit our website at:
www.cosimobooks.com

The Concept of Nature was originally published in 1920.

Cover design by www.kerndesign.net

ISBN: 978-1-60206-213-9

It is impossible to meditate on time
and the mystery of the creative passage
of nature without an overwhelming emotion
at the limitations of human intelligence.

—from Chapter III: "Time"

CONTENTS

THE CONCEPT OF NATURE

CHAPTER I

NATURE AND THOUGHT

THE subject-matter of the Tarner lectures is defined by the founder to be 'the Philosophy of the Sciences and the Relations or Want of Relations between the different Departments of Knowledge.' It is fitting at the first lecture of this new foundation to dwell for a few moments on the intentions of the donor as expressed in this definition; and I do so the more willingly as I shall thereby be enabled to introduce the topics to which the present course is to be devoted.

We are justified, I think, in taking the second clause of the definition as in part explanatory of the earlier clause. What is the philosophy of the sciences? It is not a bad answer to say that it is the study of the relations between the different departments of knowledge. Then with admirable solicitude for the freedom of learning there is inserted in the definition after the word 'relations' the phrase 'or want of relations.' A disproof of relations between sciences would in itself constitute a philosophy of the sciences. But we could not dispense either with the earlier or the later clause. It is not every relation between sciences which enters into their philosophy. For example biology and physics are connected by the use of the microscope. Still, I may safely assert that a technical description of the uses of the microscope in biology is not part of the philosophy of the sciences. Again, you cannot abandon the later clause of the definition; namely that referring to the relations between the sciences, without abandoning the explicit reference to an ideal in the absence of which philosophy must languish from lack of intrinsic interest. That ideal is the attainment of some unifying concept which will set in assigned relationships within itself all that there is for knowledge, for feeling, and for emotion. That far off ideal is the motive power of philosophic research; and claims allegiance even as you expel it. The philosophic pluralist is a strict logician; the Hegelian thrives on contradictions by the help of his absolute; the Mohammedan divine bows before the creative will of Allah; and the pragmatist will swallow anything so long as it 'works.'

The mention of these vast systems and of the age-long controversies from which they spring, warns us to concentrate. Our task is the simpler one of the philosophy of the sciences. Now a science has already a certain unity which is the very reason why that body of knowledge has been instinctively recognised as forming a science. The philosophy of a science is the endeavour to express explicitly those unifying characteristics which pervade that complex of thoughts and make it to be a science. The philosophy of the sciences—conceived as one subject—is the endeavour to exhibit all sciences as one science, or—in case of defeat—the disproof of such a possibility.

Again I will make a further simplification, and confine attention to the natural sciences, that is, to the sciences whose subject-matter is nature. By postulating a common subject-matter for this group of sciences a unifying philosophy of natural science has been thereby presupposed.

What do we mean by nature? We have to discuss the philosophy of natural science. Natural science is the science of nature. But—What is nature?

Nature is that which we observe in perception through the senses. In this sense-perception we are aware of something which is not thought and which is self-contained for thought. This property of being self-contained for thought lies at the base of natural science. It means that nature can be thought of as a closed system whose mutual relations do not require the expression of the fact that they are thought about.

Thus in a sense nature is independent of thought. By this statement no metaphysical pronouncement is intended. What I mean is that we can think about nature without thinking about thought. I shall say that then we are thinking 'homogeneously' about nature.

Of course it is possible to think of nature in conjunction with thought about the fact that nature is thought about. In such a case I shall say that we are thinking 'heterogeneously' about nature. In fact during the last few minutes we have been thinking heterogeneously about nature. Natural science is exclusively concerned with homogeneous thoughts about nature.

But sense-perception has in it an element which is not thought. It is a difficult psychological question whether sense-perception involves thought; and if it does involve thought, what is the kind of thought which it necessarily involves. Note that it has been stated above that sense-perception is an awareness of something which is not thought. Namely, nature is not thought. But this is a different question, namely that the fact of sense-perception has a factor which is not thought. I call this factor 'sense-awareness.' Accordingly the doctrine that natural science is exclusively concerned with homogeneous thoughts about nature does not immediately carry with it the conclusion that natural science is not concerned with sense-awareness.

However, I do assert this further statement; namely, that though natural science is concerned with nature which is the terminus of sense-perception, it is not concerned with the sense-awareness itself.

I repeat the main line of this argument, and expand it in certain directions.

Thought about nature is different from the sense-perception of nature. Hence the fact of sense-perception has an ingredient or factor which is not thought. I call this ingredient sense-awareness. It is indifferent to my argument whether sense-perception has or has not thought as another ingredient. If sense-perception does not involve thought, then sense-awareness and sense-perception are identical. But the something perceived is perceived as an entity which is the terminus of the sense-awareness, something which for thought is beyond the fact of that sense-awareness. Also the something perceived certainly does not contain other sense-awarenesses which are different from the sense-awareness which is an ingredient in that perception. Accordingly nature as disclosed in sense-perception is self-contained as against sense-awareness, in addition to being self-contained as against thought. I will also express this self-containedness of nature by saying that nature is closed to mind.

This closure of nature does not carry with it any metaphysical doctrine of the disjunction of nature and mind. It means that in sense-perception nature is disclosed as a complex of entities whose mutual relations are expressible in thought without reference to mind, that is, without reference either to sense-awareness or to thought. Furthermore, I do not wish to be understood as implying that sense-awareness and thought are the only activities which are to be ascribed to mind. Also I am not denying that there are relations of natural entities to mind or minds other than being the termini of the sense-awarenesses of minds. Accordingly I will extend the meaning of the terms 'homogeneous thoughts' and 'heterogeneous thoughts' which have already been introduced. We are thinking 'homogeneously' about nature when we are thinking about it without thinking about thought or about sense-awareness, and we are thinking 'heterogeneously' about nature when we are thinking about it in conjunction with thinking either about thought or about sense-awareness or about both.

I also take the homogeneity of thought about nature as excluding any reference to moral or aesthetic values whose apprehension is vivid in proportion to self-conscious activity. The values of nature are perhaps the key to the metaphysical synthesis of existence. But such a synthesis is exactly what I am not attempting. I am concerned exclusively with the generalisations of widest scope which can be effected respecting that which is known to us as the direct deliverance of sense-awareness.

I have said that nature is disclosed in sense-perception as a complex of entities. It is worth considering what we mean by an entity in this connexion. 'Entity' is simply the Latin equivalent for 'thing' unless some arbitrary distinction is drawn between the words for technical purposes. All thought has to be about things. We can gain some idea of this necessity of things for thought by examination of the structure of a proposition.

Let us suppose that a proposition is being communicated by an expositor to a recipient. Such a proposition is composed of phrases; some of these phrases may be demonstrative and others may be descriptive.

By a demonstrative phrase I mean a phrase which makes the recipient aware of an entity in a way which is independent of the particular demonstrative phrase. You will understand that I am here using 'demonstration' in the non-logical sense, namely in the sense in which a lecturer demonstrates by the aid of a frog and a microscope the circulation of the blood for an elementary class of medical students. I will call such demonstration 'speculative' demonstration, remembering Hamlet's use of the word 'speculation' when he says,

> There is no speculation in those eyes.

Thus a demonstrative phrase demonstrates an entity speculatively. It may happen that the expositor has meant some other entity—namely, the phrase demonstrates to him an entity which is diverse from the entity which it demonstrates to the recipient. In that case there is confusion; for there are two diverse propositions, namely the proposition for the expositor and the proposition for the recipient. I put this possibility aside as irrelevant for our discussion, though in practice it may be difficult for two persons to concur in the consideration of exactly the same proposition, or even for one person to have determined exactly the proposition which he is considering.

Again the demonstrative phrase may fail to demonstrate any entity. In that case there is no proposition for the recipient. I think that we may assume (perhaps rashly) that the expositor knows what he means.

A demonstrative phrase is a gesture. It is not itself a constituent of the proposition, but the entity which it demonstrates is such a constituent. You may quarrel with a demonstrative phrase as in some way obnoxious to you; but if it demonstrates the right entity, the proposition is unaffected though your taste may be offended. This suggestiveness of the phraseology is part of the literary quality of the sentence which conveys the proposition. This is because a sentence directly conveys one proposition, while in its phraseology it suggests a penumbra of other propositions charged with emotional value. We are now talking of the one proposition directly conveyed in any phraseology.

This doctrine is obscured by the fact that in most cases what is in form a mere part of the demonstrative gesture is in fact a part of the proposition which it is desired directly to convey. In such a case we will call the phraseology of the proposition elliptical. In ordinary intercourse the phraseology of nearly all propositions is elliptical.

Let us take some examples. Suppose that the expositor is in London, say in Regent's Park and in Bedford College, the great women's college which is situated in that park. He is speaking in the college hall and he says,

'This college building is commodious.'

The phrase 'this college building' is a demonstrative phrase. Now suppose the recipient answers,

'This is not a college building, it is the lion-house in the Zoo.'

Then, provided that the expositor's original proposition has not been couched in elliptical phraseology, the expositor sticks to his original proposition when he replies,

'Anyhow, *it* is commodious.'

Note that the recipient's answer accepts the speculative demonstration of the phrase 'This college building.' He does not say, 'What do you mean?' He accepts the phrase as demonstrating an entity, but declares that same entity to be the lion-house in the Zoo. In his reply, the expositor in his turn recognises the success of his original gesture as a speculative demonstration, and waives the question of the suitability of its mode of suggestiveness with an 'anyhow.' But he is now in a position to repeat the original proposition with the aid of a demonstrative gesture robbed of any suggestiveness, suitable or unsuitable, by saying,

'*It* is commodious.'

The '*it*' of this final statement presupposes that thought has seized on the entity as a bare objective for consideration.

We confine ourselves to entities disclosed in sense-awareness. The entity is so disclosed as a relatum in the complex which is nature. It dawns on an observer because of its relations; but it is an objective for thought in its own bare individuality. Thought cannot proceed otherwise; namely, it cannot proceed without the ideal bare 'it' which is speculatively demonstrated. This setting up of the entity as a bare objective does not ascribe to it an existence apart from the complex in which it has been found by sense-perception. The 'it' for thought is essentially a relatum for sense-awareness.

The chances are that the dialogue as to the college building takes another form. Whatever the expositor originally meant, he almost certainly now takes his former statement as couched in elliptical phraseology, and assumes that he was meaning,

'This is a college building and is commodious.'

Here the demonstrative phrase or the gesture, which demonstrates the 'it' which is commodious, has now been reduced to 'this'; and the attenuated phrase, under the circumstances in which it is uttered, is sufficient for the purpose of correct demonstration. This brings out the point that the verbal form is never the whole phraseology of the proposition; this phraseology also includes the general circumstances of its production. Thus the aim of a demonstrative phrase is to exhibit a definite 'it' as a bare objective for thought; but the *modus operandi* of a demonstrative phrase is to produce an awareness of the entity as a particular relatum in an auxiliary complex, chosen merely for the sake of the speculative demonstration and irrelevant to the proposition. For example, in the above dialogue, colleges and buildings, as related to the 'it' speculatively demonstrated by the phrase 'this college building,' set that 'it' in an auxiliary complex which is irrelevant to the proposition

'It is commodious.'

Of course in language every phrase is invariably highly elliptical. Accordingly the sentence

'This college building is commodious'

means probably

'This college building is commodious as a college building.'

But it will be found that in the above discussion we can replace 'commodious' by 'commodious as a college building' without altering our conclusion; though we can guess that the recipient, who thought he was in the lion-house of the Zoo, would be less likely to assent to

'Anyhow, it is commodious as a college building.'

A more obvious instance of elliptical phraseology arises if the expositor should address the recipient with the remark,

'That criminal is your friend.'

The recipient might answer,

'He is my friend and you are insulting.'

Here the recipient assumes that the phrase 'That criminal' is elliptical and not merely demonstrative. In fact, pure demonstration is impossible though it is the ideal of thought. This practical impossibility of pure demonstration is a difficulty which arises in the communication of thought and in the retention of thought. Namely, a proposition about a particular factor in nature can neither be expressed to others nor retained for repeated consideration without the aid of auxiliary complexes which are irrelevant to it.

I now pass to descriptive phrases. The expositor says,

'A college in Regent's Park is commodious.'

The recipient knows Regent's Park well. The phrase 'A college in Regent's Park' is descriptive for him. If its phraseology is not elliptical, which in ordinary life it certainly will be in some way or other, this proposition simply means,

'There is an entity which is a college building in Regent's Park and is commodious.'

If the recipient rejoins,

'The lion-house in the Zoo is the only commodious building in Regent's Park,'

he now contradicts the expositor, on the assumption that a lion-house in a Zoo is not a college building.

Thus whereas in the first dialogue the recipient merely quarrelled with the expositor without contradicting him, in this dialogue he contradicts him. Thus a descriptive phrase is part of the proposition which it helps to express, whereas a demonstrative phrase is not part of the proposition which it helps to express.

Again the expositor might be standing in Green Park —where there are no college buildings—and say,

'This college building is commodious.'

Probably no proposition will be received by the recipient because the demonstrative phrase,

'This college building'

has failed to demonstrate owing to the absence of the background of sense-awareness which it presupposes.

But if the expositor had said,

'A college building in Green Park is commodious,'

the recipient would have received a proposition, but a false one.

Language is usually ambiguous and it is rash to make general assertions as to its meanings. But phrases which commence with 'this' or 'that' are usually demonstrative, whereas phrases which commence with 'the' or 'a' are often descriptive. In studying the theory of propositional expression it is important to remember the wide difference between the analogous modest words 'this' and 'that' on the one hand and 'a' and 'the' on the other hand. The sentence

'The college building in Regent's Park is commodious'

means, according to the analysis first made by Bertrand Russell, the proposition,

'There is an entity which (i) is a college building in Regent's Park and (ii) is commodious and (iii) is such that any college building in Regent's Park is identical with it.'

The descriptive character of the phrase 'The college building in Regent's Park' is thus evident. Also the proposition is denied by the denial of any one of its three component clauses or by the denial of any combination of the component clauses. If we had substituted 'Green Park' for 'Regent's Park' a false proposition would have resulted. Also the erection of a second college in Regent's Park would make the proposition false, though in ordinary life common sense would politely treat it as merely ambiguous.

'The Iliad' for a classical scholar is usually a demonstrative phrase; for it demonstrates to him a well-known poem. But for the majority of mankind the phrase is descriptive, namely, it is synonymous with 'The poem named "the Iliad".'

Names may be either demonstrative or descriptive phrases. For example 'Homer' is for us a descriptive phrase, namely, the word with some slight difference in suggestiveness means 'The man who wrote the Iliad.'

This discussion illustrates that thought places before itself bare objectives, entities as we call them, which the thinking clothes by expressing their mutual relations. Sense-awareness discloses fact with factors which are the entities for thought. The separate distinction of an entity in thought is not a metaphysical assertion, but a method of procedure necessary for the finite expression of individual propositions. Apart from entities there could be no finite truths; they are the means by which the infinitude of irrelevance is kept out of thought.

To sum up: the termini for thought are entities,

primarily with bare individuality, secondarily with properties and relations ascribed to them in the procedure of thought; the termini for sense-awareness are factors in the fact of nature, primarily relata and only secondarily discriminated as distinct individualities.

No characteristic of nature which is immediately posited for knowledge by sense-awareness can be explained. It is impenetrable by thought, in the sense that its peculiar essential character which enters into experience by sense-awareness is for thought merely the guardian of its individuality as a bare entity. Thus for thought 'red' is merely a definite entity, though for awareness 'red' has the content of its individuality. The transition from the 'red' of awareness to the 'red' of thought is accompanied by a definite loss of content, namely by the transition from the factor 'red' to the entity 'red.' This loss in the transition to thought is compensated by the fact that thought is communicable whereas sense-awareness is incommunicable.

Thus there are three components in our knowledge of nature, namely, fact, factors, and entities. Fact is the undifferentiated terminus of sense-awareness; factors are termini of sense-awareness, differentiated as elements of fact; entities are factors in their function as the termini of thought. The entities thus spoken of are natural entities. Thought is wider than nature, so that there are entities for thought which are not natural entities.

When we speak of nature as a complex of related entities, the 'complex' is fact as an entity for thought, to whose bare individuality is ascribed the property of embracing in its complexity the natural entities. It is our business to analyse this conception and in the course of the analysis space and time should appear. Evidently the relations holding between natural entities are themselves natural entities, namely they are also factors of fact, there for sense-awareness. Accordingly the structure of the natural complex can never be completed in thought, just as the factors of fact can never be exhausted in sense-awareness. Unexhaustiveness is an essential character of our knowledge of nature. Also nature does not exhaust the matter for thought, namely there are thoughts which would not occur in any homogeneous thinking about nature.

The question as to whether sense-perception involves thought is largely verbal. If sense-perception involves a cognition of individuality abstracted from the actual position of the entity as a factor in fact, then it undoubtedly does involve thought. But if it is conceived as sense-awareness of a factor in fact competent to evoke emotion and purposeful action without further cognition, then it does not involve thought. In such a case the terminus of the sense-awareness is something for mind, but nothing for thought. The sense-perception of some lower forms of life may be conjectured to approximate to this character habitually. Also occasionally our own sense-perception in moments when thought-activity has been lulled to quiescence is not far off the attainment of this ideal limit.

The process of discrimination in sense-awareness has two distinct sides. There is the discrimination of fact into parts, and the discrimination of any part of fact as exhibiting relations to entities which are not parts of fact though they are ingredients in it. Namely the immediate fact for awareness is the whole occurrence of nature. It is nature as an event present for sense-awareness, and essentially passing. There is no holding nature still and looking at it. We cannot redouble our efforts to improve our knowledge of the terminus of our present sense-awareness; it is our subsequent opportunity in subsequent sense-awareness which gains the benefit of our good resolution. Thus the ultimate fact for sense-awareness is an event. This whole event is discriminated by us into partial events. We are aware of an event which is our bodily life, of an event which is the course of nature within this room, and of a vaguely perceived aggregate of other partial events. This is the discrimination in sense-awareness of fact into parts.

I shall use the term 'part' in the arbitrarily limited sense of an event which is part of the whole fact disclosed in awareness.

Sense-awareness also yields to us other factors in nature which are not events. For example, sky-blue is seen as situated in a certain event. This relation of situation requires further discussion which is postponed to a later lecture. My present point is that sky-blue is found in nature with a definite implication in events, but is not an event itself. Accordingly in addition to events, there are other factors in nature directly disclosed to us in sense-awareness. The conception in thought of all the factors in nature as distinct entities with definite natural relations is what I have in another place[1] called the 'diversification of nature.'

There is one general conclusion to be drawn from the foregoing discussion. It is that the first task of a philosophy of science should be some general classification of the entities disclosed to us in sense-perception.

Among the examples of entities in addition to 'events' which we have used for the purpose of illustration are

[1] Cf. *Enquiry*.

the buildings of Bedford College, Homer, and sky-blue. Evidently these are very different sorts of things; and it is likely that statements which are made about one kind of entity will not be true about other kinds. If human thought proceeded with the orderly method which abstract logic would suggest to it, we might go further and say that a classification of natural entities should be the first step in science itself. Perhaps you will be inclined to reply that this classification has already been effected, and that science is concerned with the adventures of material entities in space and time.

The history of the doctrine of matter has yet to be written. It is the history of the influence of Greek philosophy on science. That influence has issued in one long misconception of the metaphysical status of natural entities. The entity has been separated from the factor which is the terminus of sense-awareness. It has become the substratum for that factor, and the factor has been degraded into an attribute of the entity. In this way a distinction has been imported into nature which is in truth no distinction at all. A natural entity is merely a factor of fact, considered in itself. Its disconnexion from the complex of fact is a mere abstraction. It is not the substratum of the factor, but the very factor itself as bared in thought. Thus what is a mere procedure of mind in the translation of sense-awareness into discursive knowledge has been transmuted into a fundamental character of nature. In this way matter has emerged as being the metaphysical substratum of its properties, and the course of nature is interpreted as the history of matter.

Plato and Aristotle found Greek thought preoccupied with the quest for the simple substances in terms of which the course of events could be expressed. We may formulate this state of mind in the question, What is nature made of? The answers which their genius gave to this question, and more particularly the concepts which underlay the terms in which they framed their answers, have determined the unquestioned presuppositions as to time, space and matter which have reigned in science.

In Plato the forms of thought are more fluid than in Aristotle, and therefore, as I venture to think, the more valuable. Their importance consists in the evidence they yield of cultivated thought about nature before it had been forced into a uniform mould by the long tradition of scientific philosophy. For example in the *Timaeus* there is a presupposition, somewhat vaguely expressed, of a distinction between the general becoming of nature and the measurable time of nature. In a later lecture I have to distinguish between what I call the passage of nature and particular time-systems which exhibit certain characteristics of that passage. I will not go so far as to claim Plato in direct support of this doctrine, but I do think that the sections of the *Timaeus* which deal with time become clearer if my distinction is admitted.

This is however a digression. I am now concerned with the origin of the scientific doctrine of matter in Greek thought. In the *Timaeus* Plato asserts that nature is made of fire and earth with air and water as intermediate between them, so that 'as fire is to air so is air to water, and as air is to water so is water to earth.' He also suggests a molecular hypothesis for these four elements. In this hypothesis everything depends on the shape of the atoms; for earth it is cubical and for fire it is pyramidal. To-day physicists are again discussing the structure of the atom, and its shape is no slight factor in that structure. Plato's guesses read much more fantastically than does Aristotle's systematic analysis; but in some ways they are more valuable. The main outline of his ideas is comparable with that of modern science. It embodies concepts which any theory of natural philosophy must retain and in some sense must explain. Aristotle asked the fundamental question, What do we mean by 'substance'? Here the reaction between his philosophy and his logic worked very unfortunately. In his logic, the fundamental type of affirmative proposition is the attribution of a predicate to a subject. Accordingly, amid the many current uses of the term 'substance' which he analyses, he emphasises its meaning as 'the ultimate substratum which is no longer predicated of anything else.'

The unquestioned acceptance of the Aristotelian logic has led to an ingrained tendency to postulate a substratum for whatever is disclosed in sense-awareness, namely, to look below what we are aware of for the substance in the sense of the 'concrete thing.' This is the origin of the modern scientific concept of matter and of ether, namely they are the outcome of this insistent habit of postulation.

Accordingly ether has been invented by modern science as the substratum of the events which are spread through space and time beyond the reach of ordinary ponderable matter. Personally, I think that predication is a muddled notion confusing many different relations under a convenient common form of speech. For example, I hold that the relation of green to a blade of grass is entirely different from the relation of green to the event which is the life history of that blade for some short period, and is different from the relation of the blade to that event. In a sense I call the event the situation of the green, and in another sense it is the situation of the blade. Thus in one sense the blade is a character or property which can be predicated of the situation, and in another sense the green is a character or property of the same event which is also its situation. In this way the predication of properties veils radically different relations between entities.

Accordingly 'substance,' which is a correlative term to 'predication,' shares in the ambiguity. If we are to look for substance anywhere, I should find it in events which are in some sense the ultimate substance of nature.

Matter, in its modern scientific sense, is a return to the Ionian effort to find in space and time some stuff which composes nature. It has a more refined signification than the early guesses at earth and water by reason of a certain vague association with the Aristotelian idea of substance.

Earth, water, air, fire, and matter, and finally ether are related in direct succession so far as concerns their postulated characters of ultimate substrata of nature. They bear witness to the undying vitality of Greek philosophy in its search for the ultimate entities which are the factors of the fact disclosed in sense-awareness. This search is the origin of science.

The succession of ideas starting from the crude guesses of the early Ionian thinkers and ending in the nineteenth century ether reminds us that the scientific doctrine of matter is really a hybrid through which philosophy passed on its way to the refined Aristotelian concept of substance and to which science returned as it reacted against philosophic abstractions. Earth, fire, and water in the Ionic philosophy and the shaped elements in the *Timaeus* are comparable to the matter and ether of modern scientific doctrine. But substance represents the final philosophic concept of the substratum which underlies any attribute. Matter (in the scientific sense) is already in space and time. Thus matter represents the refusal to think away spatial and temporal characteristics and to arrive at the bare concept of an individual entity. It is this refusal which has caused the muddle of importing the mere procedure of thought into the fact of nature. The entity, bared of all characteristics except those of space and time, has acquired a physical status as the ultimate texture of nature; so that the course of nature is conceived as being merely the fortunes of matter in its adventure through space.

Thus the origin of the doctrine of matter is the outcome of uncritical acceptance of space and time as external conditions for natural existence. By this I do not mean that any doubt should be thrown on facts of space and time as ingredients in nature. What I do mean is 'the unconscious presupposition of space and time as being that within which nature is set.' This is exactly the sort of presupposition which tinges thought in any reaction against the subtlety of philosophical criticism. My theory of the formation of the scientific doctrine of matter is that first philosophy illegitimately transformed the bare entity, which is simply an abstraction necessary for the method of thought, into the metaphysical substratum of these factors in nature which in various senses are assigned to entities as their attributes; and that, as a second step, scientists (including philosophers who were scientists) in conscious or unconscious ignoration of philosophy presupposed this substratum, *qua* substratum for attributes, as nevertheless in time and space.

This is surely a muddle. The whole being of substance is as a substratum for attributes. Thus time and space should be attributes of the substance. This they palpably are not, if the matter be the substance of nature, since it is impossible to express spatio-temporal truths without having recourse to relations involving relata other than bits of matter. I waive this point however, and come to another. It is not the substance which is in space, but the attributes. What we find in space are the red of the rose and the smell of the jasmine and the noise of cannon. We have all told our dentists where our toothache is. Thus space is not a relation between substances, but between attributes.

Thus even if you admit that the adherents of substance can be allowed to conceive substance as matter, it is a fraud to slip substance into space on the plea that space expresses relations between substances. On the face of it space has nothing to do with substances, but only with their attributes. What I mean is, that if you choose—as I think wrongly—to construe our experience of nature as an awareness of the attributes of substances, we are by this theory precluded from finding any analogous direct relations between substances as disclosed in our experience. What we do find are relations between the attributes of substances. Thus if matter is looked on as substance in space, the space in which it finds itself has very little to do with the space of our experience.

The above argument has been expressed in terms of the relational theory of space. But if space be absolute—namely, if it have a being independent of things in it—the course of the argument is hardly changed. For things in space must have a certain fundamental relation to space which we will call occupation. Thus the objection that it is the attributes which are observed as related to space, still holds.

The scientific doctrine of matter is held in conjunction with an absolute theory of time. The same arguments apply to the relations between matter and time as apply to the relations between space and matter. There is however (in the current philosophy) a difference in the connexions of space with matter from those of time with matter, which I will proceed to explain.

Space is not merely an ordering of material entities so that any one entity bears certain relations to other material entities. The occupation of space impresses a certain character on each material entity in itself. By reason of its occupation of space matter has extension. By reason of its extension each bit of matter is divisible into parts, and each part is a numerically distinct entity from every other such part. Accordingly it would seem that every material entity is not really one entity. It is an essential multiplicity of entities. There seems to be no stopping this dissociation of matter into multiplicities short of finding each ultimate entity occupying one individual point. This essential multiplicity of material entities is certainly not what is meant by science, nor does it correspond to anything disclosed in sense-awareness. It is absolutely necessary that at a certain stage in this dissociation of matter a halt should be called, and that the material entities thus obtained should be treated as units. The stage of arrest may be arbitrary or may be set by the characteristics of nature; but all reasoning in science ultimately drops its space-analysis and poses to itself the problem, 'Here is one material entity, what is happening to it as a unit entity?' Yet this material entity is still retaining its extension, and as thus extended is a mere multiplicity. Thus there is an essential atomic property in nature which is independent of the dissociation of extension. There is something which in itself is one, and which is more than the logical aggregate of entities occupying points within the volume which the unit occupies. Indeed we may well be sceptical as to these ultimate entities at points, and doubt whether there are any such entities at all. They have the suspicious character that we are driven to accept them by abstract logic and not by observed fact.

Time (in the current philosophy) does not exert the same disintegrating effect on matter which occupies it. If matter occupies a duration of time, the whole matter occupies every part of that duration. Thus the connexion between matter and time differs from the connexion between matter and space as expressed in current scientific philosophy. There is obviously a greater difficulty in conceiving time as the outcome of relations between different bits of matter than there is in the analogous conception of space. At an instant distinct volumes of space are occupied by distinct bits of matter. Accordingly there is so far no intrinsic difficulty in conceiving that space is merely the resultant of relations between the bits of matter. But in the one-dimensional time the same bit of matter occupies different portions of time. Accordingly time would have to be expressible in terms of the relations of a bit of matter with itself. My own view is a belief in the relational theory both of space and of time, and of disbelief in the current form of the relational theory of space which exhibits bits of matter as the relata for spatial relations. The true relata are events. The distinction which I have just pointed out between time and space in their connexion with matter makes it evident that any assimilation of time and space cannot proceed along the traditional line of taking matter as a fundamental element in space-formation.

The philosophy of nature took a wrong turn during its development by Greek thought. This erroneous presupposition is vague and fluid in Plato's *Timaeus*. The general groundwork of the thought is still uncommitted and can be construed as merely lacking due explanation and the guarding emphasis. But in Aristotle's exposition the current conceptions were hardened and made definite so as to produce a faulty analysis of the relation between the matter and the form of nature as disclosed in sense-awareness. In this phrase the term 'matter' is not used in its scientific sense.

I will conclude by guarding myself against a misapprehension. It is evident that the current doctrine of matter enshrines some fundamental law of nature. Any simple illustration will exemplify what I mean. For example, in a museum some specimen is locked securely in a glass case. It stays there for years: it loses its colour, and perhaps falls to pieces. But it is the same specimen; and the same chemical elements and the same quantities of those elements are present within the case at the end as were present at the beginning. Again the engineer and the astronomer deal with the motions of real permanences in nature. Any theory of nature which for one moment loses sight of these great basic facts of experience is simply silly. But it is permissible to point out that the scientific expression of these facts has become entangled in a maze of doubtful metaphysics; and that, when we remove the metaphysics and start afresh on an unprejudiced survey of nature, a new light is thrown on many fundamental concepts which dominate science and guide the progress of research.

CHAPTER II

THEORIES OF THE BIFURCATION OF NATURE

In my previous lecture I criticised the concept of matter as the substance whose attributes we perceive. This way of thinking of matter is, I think, the historical reason for its introduction into science, and is still the vague view of it at the background of our thoughts which makes the current scientific doctrine appear so obvious. Namely we conceive ourselves as perceiving attributes of things, and bits of matter are the things whose attributes we perceive.

In the seventeenth century the sweet simplicity of this aspect of matter received a rude shock. The transmission doctrines of science were then in process of elaboration and by the end of the century were unquestioned, though their particular forms have since been modified. The establishment of these transmission theories marks a turning point in the relation between science and philosophy. The doctrines to which I am especially alluding are the theories of light and sound. I have no doubt that the theories had been vaguely floating about before as obvious suggestions of common sense; for nothing in thought is ever completely new. But at that epoch they were systematised and made exact, and their complete consequences were ruthlessly deduced. It is the establishment of this procedure of taking the consequences seriously which marks the real discovery of a theory. Systematic doctrines of light and sound as being something proceeding from the emitting bodies were definitely established, and in particular the connexion of light with colour was laid bare by Newton.

The result completely destroyed the simplicity of the 'substance and attribute' theory of perception. What we see depends on the light entering the eye. Furthermore we do not even perceive what enters the eye. The things transmitted are waves or—as Newton thought—minute particles, and the things seen are colours. Locke met this difficulty by a theory of primary and secondary qualities. Namely, there are some attributes of the matter which we do perceive. These are the primary qualities, and there are other things which we perceive, such as colours, which are not attributes of matter, but are perceived by us as if they were such attributes. These are the secondary qualities of matter.

Why should we perceive secondary qualities? It seems an extremely unfortunate arrangement that we should perceive a lot of things that are not there. Yet this is what the theory of secondary qualities in fact comes to. There is now reigning in philosophy and in science an apathetic acquiescence in the conclusion that no coherent account can be given of nature as it is disclosed to us in sense-awareness, without dragging in its relations to mind. The modern account of nature is not, as it should be, merely an account of what the mind knows of nature; but it is also confused with an account of what nature does to the mind. The result has been disastrous both to science and to philosophy, but chiefly to philosophy. It has transformed the grand question of the relations between nature and mind into the petty form of the interaction between the human body and mind.

Berkeley's polemic against matter was based on this confusion introduced by the transmission theory of light. He advocated, rightly as I think, the abandonment of the doctrine of matter in its present form. He had however nothing to put in its place except a theory of the relation of finite minds to the divine mind.

But we are endeavouring in these lectures to limit ourselves to nature itself and not to travel beyond entities which are disclosed in sense-awareness.

Percipience in itself is taken for granted. We consider indeed conditions for percipience, but only so far as those conditions are among the disclosures of perception. We leave to metaphysics the synthesis of the knower and the known. Some further explanation and defence of this position is necessary, if the line of argument of these lectures is to be comprehensible.

The immediate thesis for discussion is that any metaphysical interpretation is an illegitimate importation into the philosophy of natural science. By a metaphysical interpretation I mean any discussion of the how (beyond nature) and of the why (beyond nature) of thought and sense-awareness. In the philosophy of science we seek the general notions which apply to nature, namely, to what we are aware of in perception. It is the philosophy of the thing perceived, and it should not be confused with the metaphysics of reality of which the scope embraces both perceiver and perceived. No perplexity concerning the object of knowledge can be solved by saying that there is a mind knowing it[1].

In other words, the ground taken is this: sense-awareness is an awareness of something. What then is the general character of that something of which we

[1] Cf. *Enquiry*, preface.

are aware? We do not ask about the percipient or about the process, but about the perceived. I emphasise this point because discussions on the philosophy of science are usually extremely metaphysical—in my opinion, to the great detriment of the subject.

The recourse to metaphysics is like throwing a match into the powder magazine. It blows up the whole arena. This is exactly what scientific philosophers do when they are driven into a corner and convicted of incoherence. They at once drag in the mind and talk of entities in the mind or out of the mind as the case may be. For natural philosophy everything perceived is in nature. We may not pick and choose. For us the red glow of the sunset should be as much part of nature as are the molecules and electric waves by which men of science would explain the phenomenon. It is for natural philosophy to analyse how these various elements of nature are connected.

In making this demand I conceive myself as adopting our immediate instinctive attitude towards perceptual knowledge which is only abandoned under the influence of theory. We are instinctively willing to believe that by due attention, more can be found in nature than that which is observed at first sight. But we will not be content with less. What we ask from the philosophy of science is some account of the coherence of things perceptively known.

This means a refusal to countenance any theory of psychic additions to the object known in perception. For example, what is given in perception is the green grass. This is an object which we know as an ingredient in nature. The theory of psychic additions would treat the greenness as a psychic addition furnished by the perceiving mind, and would leave to nature merely the molecules and the radiant energy which influence the mind towards that perception. My argument is that this dragging in of the mind as making additions of its own to the thing posited for knowledge by sense-awareness is merely a way of shirking the problem of natural philosophy. That problem is to discuss the relations *inter se* of things known, abstracted from the bare fact that they are known. Natural philosophy should never ask, what is in the mind and what is in nature. To do so is a confession that it has failed to express relations between things perceptively known, namely to express those natural relations whose expression is natural philosophy. It may be that the task is too hard for us, that the relations are too complex and too various for our apprehension, or are too trivial to be worth the trouble of exposition. It is indeed true that we have gone but a very small way in the adequate formulation of such relations. But at least do not let us endeavour to conceal failure under a theory of the byplay of the perceiving mind.

What I am essentially protesting against is the bifurcation of nature into two systems of reality, which, in so far as they are real, are real in different senses. One reality would be the entities such as electrons which are the study of speculative physics. This would be the reality which is there for knowledge; although on this theory it is never known. For what is known is the other sort of reality, which is the byplay of the mind. Thus there would be two natures, one is the conjecture and the other is the dream.

Another way of phrasing this theory which I am arguing against is to bifurcate nature into two divisions, namely into the nature apprehended in awareness and the nature which is the cause of awareness. The nature which is the fact apprehended in awareness holds within it the greenness of the trees, the song of the birds, the warmth of the sun, the hardness of the chairs, and the feel of the velvet. The nature which is the cause of awareness is the conjectured system of molecules and electrons which so affects the mind as to produce the awareness of apparent nature. The meeting point of these two natures is the mind, the causal nature being influent and the apparent nature being effluent.

There are four questions which at once suggest themselves for discussion in connexion with this bifurcation theory of nature. They concern (i) causality, (ii) time, (iii) space, and (iv) delusions. These questions are not really separable. They merely constitute four distinct starting points from which to enter upon the discussion of the theory.

Causal nature is the influence on the mind which is the cause of the effluence of apparent nature from the mind. This conception of causal nature is not to be confused with the distinct conception of one part of nature as being the cause of another part. For example, the burning of the fire and the passage of heat from it through intervening space is the cause of the body, its nerves and its brain, functioning in certain ways. But this is not an action of nature on the mind. It is an interaction within nature. The causation involved in this interaction is causation in a different sense from the influence of this system of bodily interactions within nature on the alien mind which thereupon perceives redness and warmth.

The bifurcation theory is an attempt to exhibit

natural science as an investigation of the cause of the fact of knowledge. Namely, it is an attempt to exhibit apparent nature as an effluent from the mind because of causal nature. The whole notion is partly based on the implicit assumption that the mind can only know that which it has itself produced and retains in some sense within itself, though it requires an exterior reason both as originating and as determining the character of its activity. But in considering knowledge we should wipe out all these spatial metaphors, such as 'within the mind' and 'without the mind.' Knowledge is ultimate. There can be no explanation of the 'why' of knowledge; we can only describe the 'what' of knowledge. Namely we can analyse the content and its internal relations, but we cannot explain why there is knowledge. Thus causal nature is a metaphysical chimera; though there is need of a metaphysics whose scope transcends the limitation to nature. The object of such a metaphysical science is not to explain knowledge, but exhibit in its utmost completeness our concept of reality.

However, we must admit that the causality theory of nature has its strong suit. The reason why the bifurcation of nature is always creeping back into scientific philosophy is the extreme difficulty of exhibiting the perceived redness and warmth of the fire in one system of relations with the agitated molecules of carbon and oxygen, with the radiant energy from them, and with the various functionings of the material body. Unless we produce the all-embracing relations, we are faced with a bifurcated nature; namely, warmth and redness on one side, and molecules, electrons and ether on the other side. Then the two factors are explained as being respectively the cause and the mind's reaction to the cause.

Time and space would appear to provide these all-embracing relations which the advocates of the philosophy of the unity of nature require. The perceived redness of the fire and the warmth are definitely related in time and in space to the molecules of the fire and the molecules of the body.

It is hardly more than a pardonable exaggeration to say that the determination of the meaning of nature reduces itself principally to the discussion of the character of time and the character of space. In succeeding lectures I shall explain my own view of time and space. I shall endeavour to show that they are abstractions from more concrete elements of nature, namely, from events. The discussion of the details of the process of abstraction will exhibit time and space as interconnected, and will finally lead us to the sort of connexions between their measurements which occur in the modern theory of electromagnetic relativity. But this is anticipating our subsequent line of development. At present I wish to consider how the ordinary views of time and space help, or fail to help, in unifying our conception of nature.

First, consider the absolute theories of time and space. We are to consider each, namely both time and space, to be a separate and independent system of entities, each system known to us in itself and for itself concurrently with our knowledge of the events of nature. Time is the ordered succession of durationless instants; and these instants are known to us merely as the relata in the serial relation which is the time-ordering relation, and the time-ordering relation is merely known to us as relating the instants. Namely, the relation and the instants are jointly known to us in our apprehension of time, each implying the other.

This is the absolute theory of time. Frankly, I confess that it seems to me to be very unplausible. I cannot in my own knowledge find anything corresponding to the bare time of the absolute theory. Time is known to me as an abstraction from the passage of events. The fundamental fact which renders this abstraction possible is the passing of nature, its development, its creative advance, and combined with this fact is another characteristic of nature, namely the extensive relation between events. These two facts, namely the passage of events and the extension of events over each other, are in my opinion the qualities from which time and space originate as abstractions. But this is anticipating my own later speculations.

Meanwhile, returning to the absolute theory, we are to suppose that time is known to us independently of any events in time. What happens in time occupies time. This relation of events to the time occupied, namely this relation of occupation, is a fundamental relation of nature to time. Thus the theory requires that we are aware of two fundamental relations, the time-ordering relation between instants, and the time-occupation relation between instants of time and states of nature which happen at those instants.

There are two considerations which lend powerful support to the reigning theory of absolute time. In the first place time extends beyond nature. Our thoughts are in time. Accordingly it seems impossible to derive time merely from relations between elements of nature. For in that case temporal relations could not relate thoughts. Thus, to use a metaphor, time would apparently have deeper roots in reality than has nature. For we can imagine thoughts related in time without any perception of nature. For example we can imagine one of Milton's angels with thoughts succeeding each other in time, who does not happen to have noticed that the Almighty has created space and set therein a material universe. As a matter of fact I think that Milton set space on the same absolute level as time. But that need not disturb the illustration. In the second place it is difficult to derive the true serial character of time from the relative theory. Each instant is irrevocable. It can never recur by the very character of time. But if on the relative theory an instant of time is simply the state of nature at that time, and the time-ordering relation is simply the relation between such states, then the irrevocableness of time would seem to mean that an actual state of all nature can never return. I admit it seems unlikely that there should ever be such a recurrence down to the smallest particular. But extreme unlikeliness is not the point. Our ignorance is so abysmal that our judgments of likeliness and unlikeliness of future events hardly count. The real point is that the exact recurrence of a state of nature seems merely unlikely, while the recurrence of an instant of time violates our whole concept of time-order. The instants of time which have passed, are passed, and can never be again.

Any alternative theory of time must reckon with these two considerations which are buttresses of the absolute theory. But I will not now continue their discussion.

The absolute theory of space is analogous to the corresponding theory of time, but the reasons for its maintenance are weaker. Space, on this theory, is a system of extensionless points which are the relata in space-ordering relations which can technically be combined into one relation. This relation does not arrange the points in one linear series analogously to the simple method of the time-ordering relation for instants. The essential logical characteristics of this relation from which all the properties of space spring are expressed by mathematicians in the axioms of geometry. From these axioms[1] as framed by modern mathematicians the whole science of geometry can be deduced by the strictest logical reasoning. The details of these axioms do not now concern us. The points and the relations are jointly known to us in our apprehension of space, each implying the other. What happens in space, occupies space. This relation of occupation is not usually stated for events but for objects. For example, Pompey's statue would be said to occupy space, but not the event which was the assassination of Julius Caesar. In this I think that ordinary usage is unfortunate, and I hold that the relations of events to space and to time are in all respects analogous. But here I am intruding my own opinions which are to be discussed in subsequent lectures. Thus the theory of absolute space requires that we are aware of two fundamental relations, the space-ordering relation, which holds between points, and the space-occupation relation between points of space and material objects.

This theory lacks the two main supports of the corresponding theory of absolute time. In the first place space does not extend beyond nature in the sense that time seems to do. Our thoughts do not seem to occupy space in quite the same intimate way in which they occupy time. For example, I have been thinking in a room, and

[1] Cf. (for example) *Projective Geometry* by Veblen and Young, vol. i. 1910, vol. ii. 1917, Ginn and Company, Boston, U.S.A.

to that extent my thoughts are in space. But it seems nonsense to ask how much volume of the room they occupied, whether it was a cubic foot or a cubic inch; whereas the same thoughts occupy a determinate duration of time, say, from eleven to twelve on a certain date.

Thus whereas the relations of a relative theory of time are required to relate thoughts, it does not seem so obvious that the relations of a relative theory of space are required to relate them. The connexion of thought with space seems to have a certain character of indirectness which appears to be lacking in the connexion of thought with time.

Again the irrevocableness of time does not seem to have any parallel for space. Space, on the relative theory, is the outcome of certain relations between objects commonly said to be in space; and whenever there are the objects, so related, there is the space. No difficulty seems to arise like that of the inconvenient instants of time which might conceivably turn up again when we thought that we had done with them.

The absolute theory of space is not now generally popular. The knowledge of bare space, as a system of entities known to us in itself and for itself independently of our knowledge of the events in nature, does not seem to correspond to anything in our experience. Space, like time, would appear to be an abstraction from events. According to my own theory it only differentiates itself from time at a somewhat developed stage of the abstractive process. The more usual way of expressing the relational theory of space would be to consider space as an abstraction from the relations between material objects.

Suppose now we assume absolute time and absolute

space. What bearing has this assumption on the concept of nature as bifurcated into causal nature and apparent nature? Undoubtedly the separation between the two natures is now greatly mitigated. We can provide them with two systems of relations in common; for both natures can be presumed to occupy the same space and the same time. The theory now is this: Causal events occupy certain periods of the absolute time and occupy certain positions of the absolute space. These events influence a mind which thereupon perceives certain apparent events which occupy certain periods in the absolute time and occupy certain positions of the absolute space; and the periods and positions occupied by the apparent events bear a determinate relation to the periods and positions occupied by the causal events.

Furthermore definite causal events produce for the mind definite apparent events. Delusions are apparent events which appear in temporal periods and spatial positions without the intervention of these causal events which are proper for influencing of the mind to their perception.

The whole theory is perfectly logical. In these discussions we cannot hope to drive an unsound theory to a logical contradiction. A reasoner, apart from mere slips, only involves himself in a contradiction when he is shying at a *reductio ad absurdum*. The substantial reason for rejecting a philosophical theory is the 'absurdum' to which it reduces us. In the case of the philosophy of natural science the 'absurdum' can only be that our perceptual knowledge has not the character assigned to it by the theory. If our opponent affirms that his knowledge has that character, we can only—after making doubly sure that we understand each other—agree to differ. Accordingly the first duty of an expositor in stating a theory in which he disbelieves is to exhibit it as logical. It is not there where his trouble lies.

Let me summarise the previously stated objections to this theory of nature. In the first place it seeks for the cause of the knowledge of the thing known instead of seeking for the character of the thing known: secondly it assumes a knowledge of time in itself apart from events related in time: thirdly it assumes a knowledge of space in itself apart from events related in space. There are in addition to these objections other flaws in the theory.

Some light is thrown on the artificial status of causal nature in this theory by asking, why causal nature is presumed to occupy time and space. This really raises the fundamental question as to what characteristics causal nature should have in common with apparent nature. Why—on this theory—should the cause which influences the mind to perception have any characteristics in common with the effluent apparent nature? In particular, why should it be in space? Why should it be in time? And more generally, What do we know about mind which would allow us to infer any particular characteristics of a cause which should influence mind to particular effects?

The transcendence of time beyond nature gives some slight reason for presuming that causal nature should occupy time. For if the mind occupies periods of time, there would seem to be some vague reason for assuming that influencing causes occupy the same periods of time, or at least, occupy periods which are strictly related to the mental periods. But if the mind does not occupy volumes of space, there seems to be no reason why causal nature should occupy any volumes of space. Thus space would seem to be merely apparent in the same sense as apparent nature is merely apparent. Accordingly if science is really investigating causes which operate on the mind, it would seem to be entirely on the wrong tack in presuming that the causes which it is seeking for have spatial relations. Furthermore there is nothing else in our knowledge analogous to these causes which influence the mind to perception. Accordingly, beyond the rashly presumed fact that they occupy time, there is really no ground by which we can determine any point of their character. They must remain for ever unknown.

Now I assume as an axiom that science is not a fairy tale. It is not engaged in decking out unknowable entities with arbitrary and fantastic properties. What then is it that science is doing, granting that it is effecting something of importance? My answer is that it is determining the character of things known, namely the character of apparent nature. But we may drop the term 'apparent'; for there is but one nature, namely the nature which is before us in perceptual knowledge. The characters which science discerns in nature are subtle characters, not obvious at first sight. They are relations of relations and characters of characters. But for all their subtlety they are stamped with a certain simplicity which makes their consideration essential in unravelling the complex relations between characters of more perceptive insistence.

The fact that the bifurcation of nature into causal and apparent components does not express what we mean by our knowledge is brought before us when we realise our thoughts in any discussion of the causes of our perceptions. For example, the fire is burning and we see a red coal. This is explained in science by radiant energy from the coal entering our eyes. But in seeking for such an explanation we are not asking what are the sort of occurrences which are fitted to cause a mind to see red. The chain of causation is entirely different. The mind is cut out altogether. The real question is, When red is found in nature, what else is found there also? Namely we are asking for an analysis of the accompaniments in nature of the discovery of red in nature. In a subsequent lecture I shall expand this line of thought. I simply draw attention to it here in order to point out that the wave-theory of light has not been adopted because waves are just the sort of things which ought to make a mind perceive colours. This is no part of the evidence which has ever been adduced for the wave-theory, yet on the causal theory of perception, it is really the only relevant part. In other words, science is not discussing the causes of knowledge, but the coherence of knowledge. The understanding which is sought by science is an understanding of relations within nature.

So far I have discussed the bifurcation of nature in connexion with the theories of absolute time and of absolute space. My reason has been that the introduction of the relational theories only weakens the case for bifurcation, and I wished to discuss this case on its strongest grounds.

For instance, suppose we adopt the relational theory of space. Then the space in which apparent nature is set is the expression of certain relations between the apparent objects. It is a set of apparent relations between apparent relata. Apparent nature is the dream, and the apparent relations of space are dream relations, and the space is the dream space. Similarly the space in which causal nature is set is the expression of certain relations between the causal objects. It is the expression of certain facts about the causal activity which is going on behind the scenes. Accordingly causal space belongs to a different order of reality to apparent space. Hence there is no pointwise connexion between the two and it is meaningless to say that the molecules of the grass are in any place which has a determinate spatial relation to the place occupied by the grass which we see. This conclusion is very paradoxical and makes nonsense of all scientific phraseology. The case is even worse if we admit the relativity of time. For the same arguments apply, and break up time into the dream time and causal time which belong to different orders of reality.

I have however been discussing an extreme form of the bifurcation theory. It is, as I think, the most defensible form. But its very definiteness makes it the more evidently obnoxious to criticism. The intermediate form allows that the nature we are discussing is always the nature directly known, and so far it rejects the bifurcation theory. But it holds that there are psychic additions to nature as thus known, and that these additions are in no proper sense part of nature. For example, we perceive the red billiard ball at its proper time, in its proper place, with its proper motion, with its proper hardness, and with its proper inertia. But its redness and its warmth, and the sound of the click as a cannon is made off it are psychic additions, namely, secondary qualities which are only the mind's way of perceiving nature. This is not only the vaguely prevalent theory, but is, I believe, the historical form of the bifurcation theory in so far as it is derived from philosophy. I shall call it the theory of psychic additions.

This theory of psychic additions is a sound common-sense theory which lays immense stress on the obvious reality of time, space, solidity and inertia, but distrusts the minor artistic additions of colour, warmth and sound.

The theory is the outcome of common-sense in retreat. It arose in an epoch when the transmission theories of science were being elaborated. For example, colour is the result of a transmission from the material object to the perceiver's eye; and what is thus transmitted is not colour. Thus colour is not part of the reality of the material object. Similarly for the same reason sounds evaporate from nature. Also warmth is due to the transfer of something which is not temperature. Thus we are left with spatio-temporal positions, and what I may term the 'pushiness' of the body. This lands us to eighteenth and nineteenth century materialism, namely, the belief that what is real in nature is matter, in time and in space and with inertia.

Evidently a distinction in quality has been presupposed separating off some perceptions due to touch from other perceptions. These touch-perceptions are perceptions of the real inertia, whereas the other perceptions are psychic additions which must be explained on the causal theory. This distinction is the product of an epoch in which physical science has got ahead of medical pathology and of physiology. Perceptions of push are just as much the outcome of transmission as are perceptions of colour. When colour is perceived the nerves of the body are excited in one way and transmit their message towards the brain, and when push is perceived other nerves of the body are excited in another way and transmit their message towards the brain. The message of the one set is not the conveyance of colour, and the message of the other set is not the conveyance of push. But in one case colour is perceived and in the other case the push due to the object. If you snip certain nerves, there is an end to the perception of colour; and if you snip certain other nerves, there is an end to the perception of push. It would appear therefore that any reasons which should remove colour from the reality of nature should also operate to remove inertia.

Thus the attempted bifurcation of apparent nature into two parts of which one part is both causal for its own appearance and for the appearance of the other part, which is purely apparent, fails owing to the failure to establish any fundamental distinction between our ways of knowing about the two parts of nature as thus partitioned. I am not denying that the feeling of muscular effort historically led to the formulation of the concept of force. But this historical fact does not warrant us in assigning a superior reality in nature to material inertia over colour or sound. So far as reality is concerned all our sense-perceptions are in the same boat, and must be treated on the same principle. The evenness of treatment is exactly what this compromise theory fails to achieve.

The bifurcation theory however dies hard. The reason is that there really is a difficulty to be faced in relating within the same system of entities the redness of the fire with the agitation of the molecules. In another lecture I will give my own explanation of the origin of the difficulty and of its solution.

Another favourite solution, the most attenuated form

which the bifurcation theory assumes, is to maintain that the molecules and ether of science are purely conceptual. Thus there is but one nature, namely apparent nature, and atoms and ether are merely names for logical terms in conceptual formulae of calculation.

But what is a formula of calculation? It is presumably a statement that something or other is true for natural occurrences. Take the simplest of all formulae, Two and two make four. This—so far as it applies to nature—asserts that if you take two natural entities, and then again two other natural entities, the combined class contains four natural entities. Such formulae which are true for any entities cannot result in the production of the concepts of atoms. Then again there are formulae which assert that there are entities in nature with such and such special properties, say, for example, with the properties of the atoms of hydrogen. Now if there are no such entities, I fail to see how any statements about them can apply to nature. For example, the assertion that there is green cheese in the moon cannot be a premiss in any deduction of scientific importance, unless indeed the presence of green cheese in the moon has been verified by experiment. The current answer to these objections is that, though atoms are merely conceptual, yet they are an interesting and picturesque way of saying something else which is true of nature. But surely if it is something else that you mean, for heaven's sake say it. Do away with this elaborate machinery of a conceptual nature which consists of assertions about things which don't exist in order to convey truths about things which do exist. I am maintaining the obvious position that scientific laws, if they are true, are statements about entities which we obtain knowledge of as being in nature; and that, if the entities to which the statements refer are not to be found in nature, the statements about them have no relevance to any purely natural occurrence. Thus the molecules and electrons of scientific theory are, so far as science has correctly formulated its laws, each of them factors to be found in nature. The electrons are only hypothetical in so far as we are not quite certain that the electron theory is true. But their hypothetical character does not arise from the essential nature of the theory in itself after its truth has been granted.

Thus at the end of this somewhat complex discussion, we return to the position which was affirmed at its beginning. The primary task of a philosophy of natural science is to elucidate the concept of nature, considered as one complex fact for knowledge, to exhibit the fundamental entities and the fundamental relations between entities in terms of which all laws of nature have to be stated, and to secure that the entities and relations thus exhibited are adequate for the expression of all the relations between entities which occur in nature.

The third requisite, namely that of adequacy, is the one over which all the difficulty occurs. The ultimate data of science are commonly assumed to be time, space, material, qualities of material, and relations between material objects. But data as they occur in the scientific laws do not relate all the entities which present themselves in our perception of nature. For example, the wave-theory of light is an excellent well-established theory; but unfortunately it leaves out colour as perceived. Thus the perceived redness—or, other colour—has to be cut out of nature and made into the reaction of the mind under the impulse of the actual events of nature. In other words this concept of the fundamental relations within nature is inadequate. Thus we have to bend our energies to the enunciation of adequate concepts.

But in so doing, are we not in fact endeavouring to solve a metaphysical problem? I do not think so. We are merely endeavouring to exhibit the type of relations which hold between the entities which we in fact perceive as in nature. We are not called on to make any pronouncement as to the psychological relation of subjects to objects or as to the status of either in the realm of reality. It is true that the issue of our endeavour may provide material which is relevant evidence for a discussion on that question. It can hardly fail to do so. But it is only evidence, and is not itself the metaphysical discussion. In order to make clear the character of this further discussion which is out of our ken, I will set before you two quotations. One is from Schelling and I extract the quotation from the work of the Russian philosopher Lossky which has recently been so excellently translated into English[1]—'In the "Philosophy of Nature" I considered the subject-object called nature in its activity of self-constructing. In order to understand it, we must rise to an intellectual intuition of nature. The empiricist does not rise thereto, and for this reason in all his explanations it is always *he himself* that proves to be constructing nature. It is no wonder, then, that his construction and that which was to be constructed so seldom coincide. A *Natur-philosoph* raises nature to independence, and makes it construct itself, and he never feels, therefore, the necessity of opposing nature

[1] *The Intuitive Basis of Knowledge*, by N. O. Lossky, transl. by Mrs Duddington, Macmillan and Co., 1919.

as constructed (*i.e.* as experience) to real nature, or of correcting the one by means of the other.'

The other quotation is from a paper read by the Dean of St Paul's before the Aristotelian Society in May of 1919. Dr Inge's paper is entitled 'Platonism and Human Immortality,' and in it there occurs the following statement: 'To sum up. The Platonic doctrine of immortality rests on the *independence* of the spiritual world. The spiritual world is not a world of unrealised ideals, over against a real world of unspiritual fact. It is, on the contrary, the real world, of which we have a true though very incomplete knowledge, over against a world of common experience which, as a complete whole, is not real, since it is compacted out of miscellaneous data, not all on the same level, by the help of the imagination. There is no world corresponding to the world of our common experience. Nature makes abstractions for us, deciding what range of vibrations we are to see and hear, what things we are to notice and remember.'

I have cited these statements because both of them deal with topics which, though they lie outside the range of our discussion, are always being confused with it. The reason is that they lie proximate to our field of thought, and are topics which are of burning interest to the metaphysically minded. It is difficult for a philosopher to realise that anyone really is confining his discussion within the limits that I have set before you. The boundary is set up just where he is beginning to get excited. But I submit to you that among the necessary prolegomena for philosophy and for natural science is a thorough understanding of the types of entities, and types of relations among those entities, which are disclosed to us in our perceptions of nature.

CHAPTER III

TIME

THE two previous lectures of this course have been mainly critical. In the present lecture I propose to enter upon a survey of the kinds of entities which are posited for knowledge in sense-awareness. My purpose is to investigate the sorts of relations which these entities of various kinds can bear to each other. A classification of natural entities is the beginning of natural philosophy. To-day we commence with the consideration of Time.

In the first place there is posited for us a general fact: namely, something is going on; there is an occurrence for definition.

This general fact at once yields for our apprehension two factors, which I will name, the 'discerned' and the 'discernible.' The discerned is comprised of those elements of the general fact which are discriminated with their own individual peculiarities. It is the field directly perceived. But the entities of this field have relations to other entities which are not particularly discriminated in this individual way. These other entities are known merely as the relata in relation to the entities of the discerned field. Such an entity is merely a 'something' which has such-and-such definite relations to some definite entity or entities in the discerned field. As being thus related, they are—owing to the particular character of these relations—known as elements of the general fact which is going on. But we are not aware of them except as entities fulfilling the functions of relata in these relations.

Thus the complete general fact, posited as occurring, comprises both sets of entities, namely the entities perceived in their own individuality and other entities merely apprehended as relata without further definition. This complete general fact is the discernible and it comprises the discerned. The discernible is all nature as disclosed in that sense-awareness, and extends beyond and comprises all of nature as actually discriminated or discerned in that sense-awareness. The discerning or discrimination of nature is a peculiar awareness of special factors in nature in respect to their peculiar characters. But the factors in nature of which we have this peculiar sense-awareness are known as not comprising all the factors which together form the whole complex of related entities within the general fact there for discernment. This peculiarity of knowledge is what I call its unexhaustive character. This character may be metaphorically described by the statement that nature as perceived always has a ragged edge. For example, there is a world beyond the room to which our sight is confined known to us as completing the space-relations of the entities discerned within the room. The junction of the interior world of the room with the exterior world beyond is never sharp. Sounds and subtler factors disclosed in sense-awareness float in from the outside. Every type of sense has its own set of discriminated entities which are known to be relata in relation with entities not discriminated by that sense. For example we see something which we do not touch and we touch something which we do not see, and we have a general sense of the space-relations between the entity disclosed in sight and the entity disclosed in touch. Thus in the first place each of these two entities is known as a relatum in a general system of space-relations and in the second place the particular mutual relation of these two entities as related to each other in this general system is determined. But the general system of space-relations relating the entity discriminated by sight with that discriminated by sight is not dependent on the peculiar character of the other entity as reported by the alternative sense. For example, the space-relations of the thing seen would have necessitated an entity as a relatum in the place of the thing touched even although certain elements of its character had not been disclosed by touch. Thus apart from the touch an entity with a certain specific relation to the thing seen would have been disclosed by sense-awareness but not otherwise discriminated in respect to its individual character. An entity merely known as spatially related to some discerned entity is what we mean by the bare idea of 'place.' The concept of place marks the disclosure in sense-awareness of entities in nature known merely by their spatial relations to discerned entities. It is the disclosure of the discernible by means of its relations to the discerned.

This disclosure of an entity as a relatum without further specific discrimination of quality is the basis of our concept of significance. In the above example the thing seen was significant, in that it disclosed its spatial relations to other entities not necessarily otherwise entering into consciousness. Thus significance is relatedness, but it is relatedness with the emphasis on one end only of the relation.

For the sake of simplicity I have confined the argument to spatial relations; but the same considerations apply to temporal relations. The concept of 'period of time' marks the disclosure in sense-awareness of entities in nature known merely by their temporal relations to discerned entities. Still further, this separation of the ideas of space and time has merely been adopted for the sake of gaining simplicity of exposition by conformity to current language. What we discern is the specific character of a place through a period of time. This is what I mean by an 'event.' We discern some specific character of an event. But in discerning an event we are also aware of its significance as a relatum in the structure of events. This structure of events is the complex of events as related by the two relations of extension and cogredience. The most simple expression of the properties of this structure are to be found in our spatial and temporal relations. A discerned event is known as related in this structure to other events whose specific characters are otherwise not disclosed in that immediate awareness except so far as that they are relata within the structure.

The disclosure in sense-awareness of the structure of events classifies events into those which are discerned in respect to some further individual character and those which are not otherwise disclosed except as elements of the structure. These signified events must include events in the remote past as well as events in the future. We are aware of these as the far off periods of unbounded time. But there is another classification of events which is also inherent in sense-awareness. These are the events which share the immediacy of the immediately present discerned events. These are the events whose characters together with those of the discerned events comprise all nature present for discernment. They form the complete general fact which is all nature now present as disclosed in that sense-awareness. It is in this second classification of events that the differentiation of space from time takes its origin. The germ of space is to be found in the mutual relations of events within the immediate general fact which is all nature now discernible, namely within the one event which is the totality of present nature. The relations of other events to this totality of nature form the texture of time.

The unity of this general present fact is expressed by the concept of simultaneity. The general fact is the whole simultaneous occurrence of nature which is now for sense-awareness. This general fact is what I have called the discernible. But in future I will call it a 'duration,' meaning thereby a certain whole of nature which is limited only by the property of being a simultaneity. Further in obedience to the principle of comprising within nature the whole terminus of sense-awareness, simultaneity must not be conceived as an irrelevant mental concept imposed upon nature. Our sense-awareness posits for immediate discernment a certain whole, here called a 'duration'; thus a duration is a definite natural entity. A duration is discriminated as a complex of partial events, and the natural entities which are components of this complex are thereby said to be 'simultaneous with this duration.' Also in a derivative sense they are simultaneous with each other in respect to this duration. Thus simultaneity is a definite natural relation. The word 'duration' is perhaps unfortunate in so far as it suggests a mere abstract stretch of time. This is not what I mean. A duration is a concrete slab of nature limited by simultaneity which is an essential factor disclosed in sense-awareness.

Nature is a process. As in the case of everything directly exhibited in sense-awareness, there can be no explanation of this characteristic of nature. All that can be done is to use language which may speculatively demonstrate it, and also to express the relation of this factor in nature to other factors.

It is an exhibition of the process of nature that each duration happens and passes. The process of nature can also be termed the passage of nature. I definitely refrain at this stage from using the word 'time,' since the measurable time of science and of civilised life generally merely exhibits some aspects of the more fundamental fact of the passage of nature. I believe that in this doctrine I am in full accord with Bergson, though he uses 'time' for the fundamental fact which I call the 'passage of nature.' Also the passage of nature is exhibited equally in spatial transition as well as in temporal transition. It is in virtue of its passage that nature is always moving on. It is involved in the meaning of this property of 'moving on' that not only is any act of sense-awareness just that act and no other, but the terminus of each act is also unique and is the terminus of no other act. Sense-awareness seizes its only chance and presents for knowledge something which is for it alone.

There are two senses in which the terminus of sense-awareness is unique. It is unique for the sense-awareness of an individual mind and it is unique for the sense-awareness of all minds which are operating under natural conditions. There is an important distinction between the two cases. (i) For one mind not only is the discerned component of the general fact exhibited in any act of sense-awareness distinct from the discerned component of the general fact exhibited in any other act of sense-awareness of that mind, but the two corresponding durations which are respectively related by simultaneity to the two discerned components are necessarily distinct. This is an exhibition of the temporal passage of nature; namely, one duration has passed into the other. Thus not only is the passage of nature an essential character of nature in its *rôle* of the terminus of sense-awareness, but it is also essential for sense-awareness in itself. It is this truth which makes time appear to extend beyond nature. But what extends beyond nature to mind is not the serial and measurable time, which exhibits merely the character of passage in nature, but the quality of passage itself which is in no way measurable except so far as it obtains in nature. That is to say, 'passage' is not measurable except as it occurs in nature in connexion with extension. In passage we reach a connexion of nature with the ultimate metaphysical reality. The quality of passage in durations is a particular exhibition in nature of a quality which extends beyond nature. For example passage is a quality not only of nature, which is the thing known, but also of sense-awareness which is the procedure of knowing. Durations have all the reality that nature has, though what that may be we need not now determine. The measurableness of time is derivative from the properties of durations. So also is the serial character of time. We shall find that there are in nature competing serial time-systems derived from different families of durations. These are a peculiarity of the character of passage as it is found in nature. This character has the reality of nature, but we must not necessarily transfer natural time to extra-natural entities. (ii) For two minds, the discerned components of the general facts exhibited in their respective acts of sense-awareness must be different. For each mind, in its awareness of nature is aware of a certain complex of related natural entities in their relations to the living body as a focus. But the associated durations may be identical. Here we are touching on that character of the passage nature which issues in the spatial relations of simultaneous bodies. This possible identity of the durations in the case of the sense-awareness of distinct minds is what binds into one nature the private experiences of sentient beings. We are here considering the spatial side of the passage of nature. Passage in this aspect of it also seems to extend beyond nature to mind.

It is important to distinguish simultaneity from instantaneousness. I lay no stress on the mere current usage of the two terms. There are two concepts which I want to distinguish, and one I call simultaneity and the other instantaneousness. I hope that the words are judiciously chosen; but it really does not matter so long as I succeed in explaining my meaning. Simultaneity is the property of a group of natural elements which in some sense are components of a duration. A duration can be all nature present as the immediate fact posited by sense-awareness. A duration retains within itself the passage of nature. There are within it antecedents and consequents which are also durations which may be the complete specious presents of quicker consciousnesses. In other words a duration retains temporal thickness. Any concept of all nature as immediately known is always a concept of some duration though it may be enlarged in its temporal thickness beyond the possible specious present of any being known to us as existing within nature. Thus simultaneity is an ultimate factor in nature, immediate for sense-awareness.

Instantaneousness is a complex logical concept of a procedure in thought by which constructed logical entities are produced for the sake of the simple expression in thought of properties of nature. Instantaneousness is the concept of all nature at an instant, where an instant is conceived as deprived of all temporal extension. For example we conceive of the distribution of matter in space at an instant. This is a very useful concept in science especially in applied mathematics; but it is a very complex idea so far as concerns its connexions with the immediate facts of sense-awareness. There is no such thing as nature at an instant posited by sense-awareness. What sense-awareness delivers over for knowledge is nature through a period. Accordingly nature at an instant, since it is not itself a natural entity, must be defined in terms of genuine natural entities. Unless we do so, our science, which employs the concept of instantaneous nature, must abandon all claim to be founded upon observation.

I will use the term 'moment' to mean 'all nature at an instant.' A moment, in the sense in which the term is here used, has no temporal extension, and is in this respect to be contrasted with a duration which has such extension. What is directly yielded to our knowledge by sense-awareness is a duration. Accordingly we have now to explain how moments are derived from durations, and also to explain the purpose served by their introduction.

A moment is a limit to which we approach as we confine attention to durations of minimum extension. Natural relations among the ingredients of a duration gain in complexity as we consider durations of increasing temporal extension. Accordingly there is an approach to ideal simplicity as we approach an ideal diminution of extension.

The word 'limit' has a precise signification in the logic of number and even in the logic of non-numerical one-dimensional series. As used here it is so far a mere metaphor, and it is necessary to explain directly the concept which it is meant to indicate.

Durations can have the two-termed relational property of extending one over the other. Thus the duration which is all nature during a certain minute extends over the duration which is all nature during the 30th second of that minute. This relation of 'extending over'—'extension' as I shall call it—is a fundamental natural relation whose field comprises more than durations. It is a relation which two limited events can have to each other. Furthermore as holding between durations the relation appears to refer to the purely temporal extension. I shall however maintain that the same relation of extension lies at the base both of temporal and spatial extension. This discussion can be postponed; and for the present we are simply concerned with the relation of extension as it occurs in its temporal aspect for the limited field of durations.

The concept of extension exhibits in thought one side of the ultimate passage of nature. This relation holds because of the special character which passage assumes in nature; it is the relation which in the case of durations expresses the properties of 'passing over.' Thus the duration which was one definite minute passed over the duration which was its 30th second. The duration of the 30th second was part of the duration of the minute. I shall use the terms 'whole' and 'part' exclusively in this sense, that the 'part' is an event which is extended over by the other event which is the 'whole.' Thus in my nomenclature 'whole' and 'part' refer exclusively to this fundamental relation of extension; and accordingly in this technical usage only events can be either wholes or parts.

The continuity of nature arises from extension. Every event extends over other events, and every event is extended over by other events. Thus in the special case of durations which are now the only events directly under consideration, every duration is part of other durations; and every duration has other durations which are parts of it. Accordingly there are no maximum durations and no minimum durations. Thus there is no atomic structure of durations, and the perfect definition of a duration, so as to mark out its individuality and distinguish it from highly analogous durations over which it is passing, or which are passing over it, is an arbitrary postulate of thought. Sense-awareness posits durations as factors in nature but does not clearly enable thought to use it as distinguishing the separate individualities of the entities of an allied group of slightly differing durations. This is one instance of the indeterminateness of sense-awareness. Exactness is an ideal of thought, and is only realised in experience by the selection of a route of approximation.

The absence of maximum and minimum durations does not exhaust the properties of nature which make up its continuity. The passage of nature involves the existence of a family of durations. When two durations belong to the same family either one contains the other, or they overlap each other in a subordinate duration without either containing the other; or they are completely separate. The excluded case is that of durations overlapping in finite events but not containing a third duration as a common part.

It is evident that the relation of extension is transitive; namely as applied to durations, if duration *A* is part of duration *B*, and duration *B* is part of duration *C*, then *A* is part of *C*. Thus the first two cases may be combined into one and we can say that two durations which belong to the same family *either* are such that there are durations which are parts of both *or* are completely separate.

Furthermore the converse of this proposition holds; namely, if two durations have other durations which are parts of both *or* if the two durations are completely separate, then they belong to the same family.

The further characteristics of the continuity of nature—so far as durations are concerned—which has not yet been formulated arises in connexion with a family of durations. It can be stated in this way: There are durations which contain as parts any two durations of the same family. For example a week contains as parts any two of its days. It is evident that a containing duration satisfies the conditions for belonging to the same family as the two contained durations.

We are now prepared to proceed to the definition of a moment of time. Consider a set of durations all taken from the same family. Let it have the following properties: (i) of any two members of the set one contains the other as a part, and (ii) there is no duration which is a common part of every member of the set.

Now the relation of whole and part is asymmetrical; and by this I mean that if *A* is part of *B*, then *B* is not part of *A*. Also we have already noted that the relation is transitive. Accordingly we can easily see that the durations of any set with the properties just enumerated must be arranged in a one-dimensional serial order in which as we descend the series we progressively reach durations of smaller and smaller temporal extension. The series may start with any arbitrarily assumed duration of any temporal extension, but in descending the series the temporal extension progressively contracts and the successive durations are packed one within the other like the nest of boxes of a Chinese toy. But the set differs from the toy in this particular: the toy has a smallest box which forms the end box of its series; but the set of durations can have no smallest duration nor can it converge towards a duration as its limit. For the parts either of the end duration or of the limit would be parts of all the durations of the set and thus the second condition for the set would be violated.

I will call such a set of durations an 'abstractive set' of durations. It is evident that an abstractive set as we pass along it converges to the ideal of all nature with no temporal extension, namely, to the ideal of all nature at an instant. But this ideal is in fact the ideal of a nonentity. What the abstractive set is in fact doing is to guide thought to the consideration of the progressive simplicity of natural relations as we progressively diminish the temporal extension of the duration considered. Now the whole point of the procedure is that the quantitative expressions of these natural properties do converge to limits though the abstractive set does not converge to any limiting duration. The laws relating these quantitative limits are the laws of nature 'at an instant,' although in truth there is no nature at an instant and there is only the abstractive set. Thus an abstractive set is effectively the entity meant when we consider an instant of time without temporal extension. It subserves all the necessary purposes of giving a definite meaning to the concept of the properties of nature at an instant. I fully agree that this concept is fundamental in the expression of physical science. The difficulty is to express our meaning in terms of the immediate deliverances of sense-awareness, and I offer the above explanation as a complete solution of the problem.

In this explanation a moment is the set of natural properties reached by a route of approximation. An abstractive series is a route of approximation. There are different routes of approximation to the same limiting set of the properties of nature. In other words there are different abstractive sets which are to be regarded as routes of approximation to the same moment. Accordingly there is a certain amount of technical detail necessary in explaining the relations of such abstractive sets with the same convergence and in guarding against possible exceptional cases. Such details are not suitable for exposition in these lectures, and I have dealt with them fully elsewhere[1].

It is more convenient for technical purposes to look on a moment as being the class of all abstractive sets of durations with the same convergence. With this definition (provided that we can successfully explain what we mean by the 'same convergence' apart from a detailed knowledge of the set of natural properties arrived at by approximation) a moment is merely a class of sets of durations whose relations of extension in respect to each other have certain definite peculiarities. We may term these connexions of the component durations the 'extrinsic' properties of a moment; the 'intrinsic' properties of the moment are the properties of nature arrived at as a limit as we proceed along any one of its abstractive sets. These are the properties of nature 'at that moment,' or 'at that instant.'

[1] Cf. *An Enquiry concerning the Principles of Natural Knowledge*, Cambridge University Press, 1919.

The durations which enter into the composition of a moment all belong to one family. Thus there is one family of moments corresponding to one family of durations. Also if we take two moments of the same family, among the durations which enter into the composition of one moment the smaller durations are completely separated from the smaller durations which enter into the composition of the other moment. Thus the two moments in their intrinsic properties must exhibit the limits of completely different states of nature. In this sense the two moments are completely separated. I will call two moments of the same family 'parallel.'

Corresponding to each duration there are two moments of the associated family of moments which are the boundary moments of that duration. A 'boundary moment' of a duration can be defined in this way. There are durations of the same family as the given duration which overlap it but are not contained in it. Consider an abstractive set of such durations. Such a set defines a moment which is just as much without the duration as within it. Such a moment is a boundary moment of the duration. Also we call upon our sense-awareness of the passage of nature to inform us that there are two such boundary moments, namely the earlier one and the later one. We will call them the initial and the final boundaries.

There are also moments of the same family such that the shorter durations in their composition are entirely separated from the given duration. Such moments will be said to lie 'outside' the given duration. Again other moments of the family are such that the shorter durations in their composition are parts of the given duration. Such moments are said to lie 'within' the given duration or to 'inhere' in it. The whole family of parallel moments is accounted for in this way by reference to any given duration of the associated family of durations. Namely, there are moments of the family which lie without the given duration, there are the two moments which are the boundary moments of the given duration, and the moments which lie within the given duration. Furthermore any two moments of the same family are the boundary moments of some one duration of the associated family of durations.

It is now possible to define the serial relation of temporal order among the moments of a family. For let A and C be any two moments of the family, these moments are the boundary moments of one duration d of the associated family, and any moment B which lies within the duration d will be said to lie between the moments A and C. Thus the three-termed relation of 'lying-between' as relating three moments A, B, and C is completely defined. Also our knowledge of the passage of nature assures us that this relation distributes the moments of the family into a serial order. I abstain from enumerating the definite properties which secure this result, I have enumerated them in my recently published book[1] to which I have already referred. Furthermore the passage of nature enables us to know that one direction along the series corresponds to passage into the future and the other direction corresponds to retrogression towards the past.

Such an ordered series of moments is what we mean by time defined as a series. Each element of the series exhibits an instantaneous state of nature. Evidently this serial time is the result of an intellectual process of

[1] Cf. *Enquiry*

abstraction. What I have done is to give precise definitions of the procedure by which the abstraction is effected. This procedure is merely a particular case of the general method which in my book I name the 'method of extensive abstraction.' This serial time is evidently not the very passage of nature itself. It exhibits some of the natural properties which flow from it. The state of nature 'at a moment' has evidently lost this ultimate quality of passage. Also the temporal series of moments only retains it as an extrinsic relation of entities and not as the outcome of the essential being of the terms of the series.

Nothing has yet been said as to the measurement of time. Such measurement does not follow from the mere serial property of time; it requires a theory of congruence which will be considered in a later lecture.

In estimating the adequacy of this definition of the temporal series as a formulation of experience it is necessary to discriminate between the crude deliverance of sense-awareness and our intellectual theories. The lapse of time is a measurable serial quantity. The whole of scientific theory depends on this assumption and any theory of time which fails to provide such a measurable series stands self-condemned as unable to account for the most salient fact in experience. Our difficulties only begin when we ask what it is that is measured. It is evidently something so fundamental in experience that we can hardly stand back from it and hold it apart so as to view it in its own proportions.

We have first to make up our minds whether time is to be found in nature or nature is to be found in time. The difficulty of the latter alternative—namely of making time prior to nature—is that time then becomes a metaphysical enigma. What sort of entities are its instants or its periods? The dissociation of time from events discloses to our immediate inspection that the attempt to set up time as an independent terminus for knowledge is like the effort to find substance in a shadow. There is time because there are happenings, and apart from happenings there is nothing.

It is necessary however to make a distinction. In some sense time extends beyond nature. It is not true that a timeless sense-awareness and a timeless thought combine to contemplate a timeful nature. Sense-awareness and thought are themselves processes as well as their termini in nature. In other words there is a passage of sense-awareness and a passage of thought. Thus the reign of the quality of passage extends beyond nature. But now the distinction arises between passage which is fundamental and the temporal series which is a logical abstraction representing some of the properties of nature. A temporal series, as we have defined it, represents merely certain properties of a family of durations—properties indeed which durations only possess because of their partaking of the character of passage, but on the other hand properties which only durations do possess. Accordingly time in the sense of a measurable temporal series is a character of nature only, and does not extend to the processes of thought and of sense-awareness except by a correlation of these processes with the temporal series implicated in their procedures.

So far the passage of nature has been considered in connexion with the passage of durations; and in this connexion it is peculiarly associated with temporal series. We must remember however that the character of passage is peculiarly associated with the extension of events, and that from this extension spatial transition arises just as much as temporal transition. The discussion of this point is reserved for a later lecture but it is necessary to remember it now that we are proceeding to discuss the application of the concept of passage beyond nature, otherwise we shall have too narrow an idea of the essence of passage.

It is necessary to dwell on the subject of sense-awareness in this connexion as an example of the way in which time concerns mind, although measurable time is a mere abstract from nature and nature is closed to mind.

Consider sense-awareness—not its terminus which is nature, but sense-awareness in itself as a procedure of mind. Sense-awareness is a relation of mind to nature. Accordingly we are now considering mind as a relatum in sense-awareness. For mind there is the immediate sense-awareness and there is memory. The distinction between memory and the present immediacy has a double bearing. On the one hand it discloses that mind is not impartially aware of all those natural durations to which it is related by awareness. Its awareness shares in the passage of nature. We can imagine a being whose awareness, conceived as his private possession, suffers no transition, although the terminus of his awareness is our own transient nature. There is no essential reason why memory should not be raised to the vividness of the present fact; and then from the side of mind, What is the difference between the present and the past? Yet with this hypothesis we can also suppose that the vivid remembrance and the present fact are posited in awareness as in their temporal serial order. Accordingly we must admit that though we can imagine that mind in the operation of sense-awareness might be free from any character of passage, yet in point of fact our experience of sense-awareness exhibits our minds as partaking in this character.

On the other hand the mere fact of memory is an escape from transience. In memory the past is present. It is not present as overleaping the temporal succession of nature, but it is present as an immediate fact for the mind. Accordingly memory is a disengagement of the mind from the mere passage of nature; for what has passed for nature has not passed for mind.

Furthermore the distinction between memory and the immediate present is not so clear as it is conventional to suppose. There is an intellectual theory of time as a moving knife-edge, exhibiting a present fact without temporal extension. This theory arises from the concept of an ideal exactitude of observation. Astronomical observations are successively refined to be exact to tenths, to hundredths, and to thousandths of seconds. But the final refinements are arrived at by a system of averaging, and even then present us with a stretch of time as a margin of error. Here error is merely a conventional term to express the fact that the character of experience does not accord with the ideal of thought. I have already explained how the concept of a moment conciliates the observed fact with this ideal; namely, there is a limiting simplicity in the quantitative expression of the properties of durations, which is arrived at by considering any one of the abstractive sets included in the moment. In other words the extrinsic character of the moment as an aggregate of durations has associated with it the intrinsic character of the moment which is the limiting expression of natural properties.

Thus the character of a moment and the ideal of exactness which it enshrines do not in any way weaken the position that the ultimate terminus of awareness is a duration with temporal thickness. This immediate duration is not clearly marked out for our apprehension. Its earlier boundary is blurred by a fading into memory, and its later boundary is blurred by an emergence from anticipation. There is no sharp distinction either between memory and the present immediacy or between the present immediacy and anticipation. The present is a wavering breadth of boundary between the two extremes. Thus our own sense-awareness with its extended present has some of the character of the sense-awareness of the imaginary being whose mind was free from passage and who contemplated all nature as an immediate fact. Our own present has its antecedents and its consequents, and for the imaginary being all nature has its antecedent and its consequent durations. Thus the only difference in this respect between us and the imaginary being is that for him all nature shares in the immediacy of our present duration.

The conclusion of this discussion is that so far as sense-awareness is concerned there is a passage of mind which is distinguishable from the passage of nature though closely allied with it. We may speculate, if we like, that this alliance of the passage of mind with the passage of nature arises from their both sharing in some ultimate character of passage which dominates all being. But this is a speculation in which we have no concern. The immediate deduction which is sufficient for us is that—so far as sense-awareness is concerned—mind is not in time or in space in the same sense in which the events of nature are in time, but that it is derivatively in time and in space by reason of the peculiar alliance of its passage with the passage of nature. Thus mind is in time and in space in a sense peculiar to itself. This has been a long discussion to arrive at a very simple and obvious conclusion. We all feel that in some sense our minds are here in this room and at this time. But it is not quite in the same sense as that in which the events of nature which are the existences of our brains have their spatial and temporal positions. The fundamental distinction to remember is that immediacy for sense-awareness is not the same as instantaneousness for nature. This last conclusion bears on the next discussion with which I will terminate this lecture. This question can be formulated thus, Can alternative temporal series be found in nature?

A few years ago such a suggestion would have been put aside as being fantastically impossible. It would have had no bearing on the science then current, and was akin to no ideas which had ever entered into the dreams of philosophy. The eighteenth and nineteenth centuries accepted as their natural philosophy a certain circle of concepts which were as rigid and definite as those of the philosophy of the middle ages, and were accepted with as little critical research. I will call this natural philosophy 'materialism.' Not only were men of science materialists, but also adherents of all schools of philosophy. The idealists only differed from the philosophic materialists on question of the alignment of nature in reference to mind. But no one had any doubt that the philosophy of nature considered in itself was of the type which I have called materialism. It is the philosophy which I have already examined in my two lectures of this course preceding the present one. It can be summarised as the belief that nature is an aggregate of material and that this material exists in some sense *at* each successive member of a one-dimensional series of extensionless instants of time. Furthermore the mutual relations of the material entities at each instant formed these entities into a spatial configuration in an unbounded space. It would seem that space—on this theory—would be as instantaneous as the instants, and that some explanation is required of the relations between the successive instantaneous spaces. The materialistic theory is however silent on this point; and the succession of instantaneous spaces is tacitly combined into one persistent space. This theory is a purely intellectual rendering of experience which has had the luck to get itself formulated at the dawn of scientific thought. It has dominated the language and the imagination of science since science flourished in Alexandria, with the result that it is now hardly possible to speak without appearing to assume its immediate obviousness.

But when it is distinctly formulated in the abstract terms in which I have just stated it, the theory is very far from obvious. The passing complex of factors which compose the fact which is the terminus of sense-awareness places before us nothing corresponding to the trinity of this natural materialism. This trinity is composed (i) of the temporal series of extensionless instants, (ii) of the aggregate of material entities, and (iii) of space which is the outcome of relations of matter.

There is a wide gap between these presuppositions of the intellectual theory of materialism and the immediate deliverances of sense-awareness. I do not question that this materialistic trinity embodies important characters of nature. But it is necessary to express these characters in terms of the facts of experience. This is exactly what in this lecture I have been endeavouring to do so far as time is concerned; and we have now come up against the question, Is there only one temporal series? The uniqueness of the temporal series is presupposed in the materialist philosophy of nature. But that philosophy is merely a theory, like the Aristotelian scientific theories so firmly believed in the middle ages. If in this lecture I have in any way succeeded in getting behind the theory to the immediate facts, the answer is not nearly so certain. The question can be transformed into this alternative form, Is there only one family of durations? In this question the meaning of a 'family of durations' has been defined earlier in this lecture. The answer is now not at all obvious. On the materialistic theory the instantaneous present is the only field for the creative activity of nature. The past is gone and the future is not yet. Thus (on this theory) the immediacy of perception is of an instantaneous present, and this unique present is the outcome of the past and the promise of the future. But we deny this immediately given instantaneous present. There is no such thing to be found in nature. As an ultimate fact it is a nonentity. What is immediate for sense-awareness is a duration. Now a duration has within itself a past and a future; and the temporal breadths of the immediate durations of sense-awareness are very indeterminate and dependent on the individual percipient. Accordingly there is no unique factor in nature which for every percipient is pre-eminently and necessarily the present. The passage of nature leaves nothing between the past and the future.

What we perceive as present is the vivid fringe of memory tinged with anticipation. This vividness lights up the discriminated field within a duration. But no assurance can thereby be given that the happenings of nature cannot be assorted into other durations of alternative families. We cannot even know that the series of immediate durations posited by the sense-awareness of one individual mind all necessarily belong to the same family of durations. There is not the slightest reason to believe that this is so. Indeed if my theory of nature be correct, it will not be the case.

The materialistic theory has all the completeness of the thought of the middle ages, which had a complete answer to everything, be it in heaven or in hell or in nature. There is a trimness about it, with its instantaneous present, its vanished past, its non-existent future, and its inert matter. This trimness is very medieval and ill accords with brute fact.

The theory which I am urging admits a greater ultimate mystery and a deeper ignorance. The past and the future meet and mingle in the ill-defined present. The passage of nature which is only another name for the creative force of existence has no narrow ledge of definite instantaneous present within which to operate. Its operative presence which is now urging nature forward must be sought for throughout the whole, in the remotest past as well as in the narrowest breadth of any present duration. Perhaps also in the unrealised future. Perhaps also in the future which might be as well as the actual future which will be. It is impossible to meditate on time and the mystery of the creative passage of nature without an overwhelming emotion at the limitations of human intelligence.

CHAPTER IV

THE METHOD OF EXTENSIVE ABSTRACTION

TO-DAY's lecture must commence with the consideration of limited events. We shall then be in a position to enter upon an investigation of the factors in nature which are represented by our conception of space.

The duration which is the immediate disclosure of our sense-awareness is discriminated into parts. There is the part which is the life of all nature within a room, and there is the part which is the life of all nature within a table in the room. These parts are limited events. They have the endurance of the present duration, and they are parts of it. But whereas a duration is an unlimited whole and in a certain limited sense is all that there is, a limited event possesses a completely defined limitation of extent which is expressed for us in spatio-temporal terms.

We are accustomed to associate an event with a certain melodramatic quality. If a man is run over, that is an event comprised within certain spatio-temporal limits. We are not accustomed to consider the endurance of the Great Pyramid throughout any definite day as an event. But the natural fact which is the Great Pyramid throughout a day, meaning thereby all nature within it, is an event of the same character as the man's accident, meaning thereby all nature with spatio-temporal limitations so as to include the man and the motor during the period when they were in contact.

We are accustomed to analyse these events into three factors, time, space, and material. In fact, we at once apply to them the concepts of the materialistic theory of nature. I do not deny the utility of this analysis for the purpose of expressing important laws of nature. What I am denying is that anyone of these factors is posited for us in sense-awareness in concrete independence. We perceive one unit factor in nature; and this factor is that something is going on then—there. For example, we perceive the going-on of the Great Pyramid in its relations to the goings-on of the surrounding Egyptian events. We are so trained, both by language and by formal teaching and by the resulting convenience, to express our thoughts in terms of this materialistic analysis that intellectually we tend to ignore the true unity of the factor really exhibited in sense-awareness. It is this unit factor, retaining in itself the passage of nature, which is the primary concrete element discriminated in nature. These primary factors are what I mean by events.

Events are the field of a two-termed relation, namely the relation of extension which was considered in the last lecture. Events are the things related by the relation of extension. If an event *A* extends over an event *B*, then *B* is 'part of' *A*, and *A* is a 'whole' of which *B* is a part. Whole and part are invariably used in these lectures in this definite sense. It follows that in reference to this relation any two events *A* and *B* may have any one of four relations to each other, namely (i) *A* may extend over *B*, or (ii) *B* may extend over *A*, or (iii) *A* and *B* may both extend over some third event *C*, but neither over the other, or (iv) *A* and *B* may be entirely separate. These alternatives can obviously be illustrated by Euler's diagrams as they appear in logical textbooks.

The continuity of nature is the continuity of events. This continuity is merely the name for the aggregate of a variety of properties of events in connexion with the relation of extension.

In the first place, this relation is transitive; secondly, every event contains other events as parts of itself; thirdly every event is a part of other events; fourthly given any two finite events there are events each of which contains both of them as parts; and fifthly there is a special relation between events which I term 'junction.'

Two events have junction when there is a third event of which both events are parts, and which is such that no part of it is separated from both of the two given events. Thus two events with junction make up exactly one event which is in a sense their sum.

Only certain pairs of events have this property. In general any event containing two events also contains parts which are separated from both events.

There is an alternative definition of the junction of two events which I have adopted in my recent book[1]. Two events have junction when there is a third event such that (i) it overlaps both events and (ii) it has no part which is separated from both the given events. If either of these alternative definitions is adopted as the definition of junction, the other definition appears as an axiom respecting the character of junction as we know it in nature. But we are not thinking of logical definition so much as the formulation of the results of direct observation. There is a certain continuity

[1] Cf. *Enquiry*.

inherent in the observed unity of an event, and these two definitions of junction are really axioms based on observation respecting the character of this continuity.

The relations of whole and part and of overlapping are particular cases of the junction of events. But it is possible for events to have junction when they are separate from each other; for example, the upper and the lower part of the Great Pyramid are divided by some imaginary horizontal plane.

The continuity which nature derives from events has been obscured by the illustrations which I have been obliged to give. For example I have taken the existence of the Great Pyramid as a fairly well-known fact to which I could safely appeal as an illustration. This is a type of event which exhibits itself to us as the situation of a recognisable object; and in the example chosen the object is so widely recognised that it has received a name. An object is an entity of a different type from an event. For example, the event which is the life of nature within the Great Pyramid yesterday and to-day is divisible into two parts, namely the Great Pyramid yesterday and the Great Pyramid to-day. But the recognisable object which is also called the Great Pyramid is the same object to-day as it was yesterday. I shall have to consider the theory of objects in another lecture.

The whole subject is invested with an unmerited air of subtlety by the fact that when the event is the situation of a well-marked object, we have no language to distinguish the event from the object. In the case of the Great Pyramid, the object is the perceived unit entity which as perceived remains self-identical throughout the ages; while the whole dance of molecules and the shifting play of the electromagnetic field are ingredients of the event. An object is in a sense out of time. It is only derivatively in time by reason of its having the relation to events which I term 'situation.' This relation of situation will require discussion in a subsequent lecture.

The point which I want to make now is that being the situation of a well-marked object is not an inherent necessity for an event. Wherever and whenever something is going on, there is an event. Furthermore 'wherever and whenever' in themselves presuppose an event, for space and time in themselves are abstractions from events. It is therefore a consequence of this doctrine that something is always going on everywhere, even in so-called empty space. This conclusion is in accord with modern physical science which presupposes the play of an electromagnetic field throughout space and time. This doctrine of science has been thrown into the materialistic form of an all-pervading ether. But the ether is evidently a mere idle concept—in the phraseology which Bacon applied to the doctrine of final causes, it is a barren virgin. Nothing is deduced from it; and the ether merely subserves the purpose of satisfying the demands of the materialistic theory. The important concept is that of the shifting facts of the fields of force. This is the concept of an ether of events which should be substituted for that of a material ether.

It requires no illustration to assure you that an event is a complex fact, and the relations between two events form an almost impenetrable maze. The clue discovered by the common sense of mankind and systematically utilised in science is what I have elsewhere[1] called the law of convergence to simplicity by diminution of extent.

If A and B are two events, and A' is part of A and B' is part of B, then in many respects the relations between the parts A' and B' will be simpler than the relations between A and B. This is the principle which presides over all attempts at exact observation.

The first outcome of the systematic use of this law has been the formulation of the abstract concepts of Time and Space. In the previous lecture I sketched how the principle was applied to obtain the time-series. I now proceed to consider how the spatial entities are obtained by the same method. The systematic procedure is identical in principle in both cases, and I have called the general type of procedure the 'method of extensive abstraction.'

You will remember that in my last lecture I defined the concept of an abstractive set of durations. This definition can be extended so as to apply to any events, limited events as well as durations. The only change that is required is the substitution of the word 'event' for the word 'duration.' Accordingly an abstractive set of events is any set of events which possesses the two properties, (i) of any two members of the set one contains the other as a part, and (ii) there is no event which is a common part of every member of the set. Such a set, as you will remember, has the properties of the Chinese toy which is a nest of boxes, one within the other, with the difference that the toy has a smallest box, while the abstractive class has neither a smallest

[1] Cf. *Organisation of Thought*, pp. 146 et seq. Williams and Norgate, 1917.

event nor does it converge to a limiting event which is not a member of the set.

Thus, so far as the abstractive sets of events are concerned, an abstractive set converges to nothing. There is the set with its members growing indefinitely smaller and smaller as we proceed in thought towards the smaller end of the series; but there is no absolute minimum of any sort which is finally reached. In fact the set is just itself and indicates nothing else in the way of events, except itself. But each event has an intrinsic character in the way of being a situation of objects and of having parts which are situations of objects and—to state the matter more generally—in the way of being a field of the life of nature. This character can be defined by quantitative expressions expressing relations between various quantities intrinsic to the event or between such quantities and other quantities intrinsic to other events. In the case of events of considerable spatio-temporal extension this set of quantitative expressions is of bewildering complexity. If e be an event, let us denote by $q(e)$ the set of quantitative expressions defining its character including its connexions with the rest of nature. Let e_1, e_2, e_3, etc. be an abstractive set, the members being so arranged that each member such as e_n extends over all the succeeding members such as e_{n+1}, e_{n+2}, and so on. Then corresponding to the series

$$e_1, e_2, e_3, \ldots, e_n, e_{n+1}, \ldots,$$

there is the series

$$q(e_1), q(e_2), q(e_3), \ldots, q(e_n), q(e_{n+1}), \ldots.$$

Call the series of events s and the series of quantitative expressions $q(s)$. The series s has no last term and no events which are contained in every member of the series. Accordingly the series of events converges to nothing. It is just itself. Also the series $q(s)$ has no last term. But the sets of homologous quantities running through the various terms of the series do converge to definite limits. For example if Q_1 be a quantitative measurement found in $q(e_1)$, and Q_2 the homologue to Q_1 to be found in $q(e_2)$, and Q_3 the homologue to Q_1 and Q_2 to be found in $q(e_3)$, and so on, then the series

$$Q_1, Q_2, Q_3, \ldots, Q_n, Q_{n+1}, \ldots,$$

though it has no last term, does in general converge to a definite limit. Accordingly there is a class of limits $l(s)$ which is the class of the limits of those members of $q(e_n)$ which have homologues throughout the series $q(s)$ as n indefinitely increases. We can represent this statement diagrammatically by using an arrow ($\rightarrow$) to mean 'converges to.' Then

$$e_1, e_2, e_3, \ldots, e_n, e_{n+1}, \ldots \rightarrow \text{nothing},$$

and

$$q(e_1), q(e_2), q(e_3), \ldots, q(e_n), q(e_{n+1}), \ldots \rightarrow l(s).$$

The mutual relations between the limits in the set $l(s)$, and also between these limits and the limits in other sets $l(s')$, $l(s'')$, ..., which arise from other abstractive sets s', s'', etc., have a peculiar simplicity.

Thus the set s does indicate an ideal simplicity of natural relations, though this simplicity is not the character of any actual event in s. We can make an approximation to such a simplicity which, as estimated numerically, is as close as we like by considering an event which is far enough down the series towards the small end. It will be noted that it is the infinite series, as it stretches away in unending succession towards the small end, which is of importance. The arbitrarily large event with which the series starts has no importance at all. We can arbitrarily exclude any set of events at the big end of an abstractive set without the loss of any important property to the set as thus modified.

I call the limiting character of natural relations which is indicated by an abstractive set, the 'intrinsic character' of the set; also the properties, connected with the relation of whole and part as concerning its members, by which an abstractive set is defined together form what I call its 'extrinsic character.' The fact that the extrinsic character of an abstractive set determines a definite intrinsic character is the reason of the importance of the precise concepts of space and time. This emergence of a definite intrinsic character from an abstractive set is the precise meaning of the law of convergence.

For example, we see a train approaching during a minute. The event which is the life of nature within that train during the minute is of great complexity and the expression of its relations and of the ingredients of its character baffles us. If we take one second of that minute, the more limited event which is thus obtained is simpler in respect to its ingredients, and shorter and shorter times such as a tenth of that second, or a hundredth, or a thousandth—so long as we have a definite rule giving a definite succession of diminishing events—give events whose ingredient characters converge to the ideal simplicity of the character of the train at a definite instant. Furthermore there are different types of such convergence to simplicity. For example, we can converge as above to the limiting character expressing nature at an instant within the whole volume of the train at that instant, or to nature at an instant within some portion of that volume—for example within the boiler of the engine—or to nature at an instant on some area of surface, or to nature at an instant on some line within the train, or to nature at an instant at some point of the train. In the last case the simple limiting characters arrived at will be expressed as densities, specific gravities, and types of material. Furthermore we need not necessarily converge to an abstraction which involves nature at an instant. We may converge to the physical ingredients of a certain point track throughout the whole minute. Accordingly there are different types of extrinsic character of convergence which lead to the approximation to different types of intrinsic characters as limits.

We now pass to the investigation of possible connexions between abstractive sets. One set may 'cover' another. I define 'covering' as follows: An abstractive set *p* covers an abstractive set *q* when every member of *p* contains as its parts some members of *q*. It is evident that if any event *e* contains as a part any member of the set *q*, then owing to the transitive property of extension every succeeding member of the small end of *q* is part of *e*. In such a case I will say that the abstractive set *q* 'inheres in' the event *e*. Thus when an abstractive set *p* covers an abstractive set *q*, the abstractive set *q* inheres in every member of *p*.

Two abstractive sets may each cover the other. When this is the case I shall call the two sets 'equal in abstractive force.' When there is no danger of misunderstanding I shall shorten this phrase by simply saying that the two abstractive sets are 'equal.' The possibility of this equality of abstractive sets arises from the fact that both sets, p and q, are infinite series towards their small ends. Thus the equality means, that given any event x belonging to p, we can always by proceeding far enough towards the small end of q find an event y which is part of x, and that then by proceeding far enough towards the small end of p we can find an event z which is part of y, and so on indefinitely.

The importance of the equality of abstractive sets arises from the assumption that the intrinsic characters of the two sets are identical. If this were not the case exact observation would be at an end.

It is evident that any two abstractive sets which are equal to a third abstractive set are equal to each other. An 'abstractive element' is the whole group of abstractive sets which are equal to any one of themselves. Thus all abstractive sets belonging to the same element are equal and converge to the same intrinsic character. Thus an abstractive element is the group of routes of approximation to a definite intrinsic character of ideal simplicity to be found as a limit among natural facts.

If an abstractive set p covers an abstractive set q, then any abstractive set belonging to the abstractive element of which p is a member will cover any abstractive set belonging to the element of which q is a member. Accordingly it is useful to stretch the meaning of the term 'covering,' and to speak of one abstractive element 'covering' another abstractive element. If we attempt in like manner to stretch the term 'equal' in the sense of 'equal in abstractive force,' it is obvious that an abstractive element can only be equal to itself. Thus an abstractive element has a unique abstractive force and is the construct from events which represents one definite intrinsic character which is arrived at as a limit by the use of the principle of convergence to simplicity by diminution of extent.

When an abstractive element *A* covers an abstractive element *B*, the intrinsic character of *A* in a sense includes the intrinsic character of *B*. It results that statements about the intrinsic character of *B* are in a sense statements about the intrinsic character of *A*; but the intrinsic character of *A* is more complex than that of *B*.

The abstractive elements form the fundamental elements of space and time, and we now turn to the consideration of the properties involved in the formation of special classes of such elements. In my last lecture I have already investigated one class of abstractive elements, namely moments. Each moment is a group of abstractive sets, and the events which are members of these sets are all members of one family of durations. The moments of one family form a temporal series; and, allowing the existence of different families of moments, there will be alternative temporal series in nature. Thus the method of extensive abstraction explains the origin of temporal series in terms of the immediate facts of experience and at the same time allows for the existence of the alternative temporal series which are demanded by the modern theory of electromagnetic relativity.

We now turn to space. The first thing to do is to get hold of the class of abstractive elements which are in some sense the points of space. Such an abstractive element must in some sense exhibit a convergence to an absolute minimum of intrinsic character. Euclid has expressed for all time the general idea of a point, as being without parts and without magnitude. It is this character of being an absolute minimum which we want to get at and to express in terms of the extrinsic characters of the abstractive sets which make up a point. Furthermore, points which are thus arrived at represent the ideal of events without any extension, though there are in fact no such entities as these ideal events. These points will not be the points of an external timeless space but of instantaneous spaces. We ultimately want to arrive at the timeless space of physical science, and also of common thought which is now tinged with the concepts of science. It will be convenient to reserve the term 'point' for these spaces when we get to them. I will therefore use the name 'event-particles' for the ideal minimum limits to events. Thus an event-particle is an abstractive element and as such is a group of abstractive sets; and a point—namely a point of timeless space—will be a class of event-particles.

Furthermore there is a separate timeless space corresponding to each separate temporal series, that is to each separate family of durations. We will come back to points in timeless spaces later. I merely mention them now that we may understand the stages of our investigation. The totality of event-particles will form a four-dimensional manifold, the extra dimension arising from time—in other words—arising from the points of a timeless space being each a class of event-particles.

The required character of the abstractive sets which form event-particles would be secured if we could define them as having the property of being covered by any abstractive set which they cover. For then any other abstractive set which an abstractive set of an event-particle covered, would be equal to it, and would therefore be a member of the same event-particle. Accordingly an event-particle could cover no other abstractive element. This is the definition which I originally proposed at a congress in Paris in 1914[1]. There is however a difficulty involved in this definition if adopted without some further addition, and I am now not satisfied with the way in which I attempted to get over that difficulty in the paper referred to.

The difficulty is this: When event-particles have once been defined it is easy to define the aggregate of event-particles forming the boundary of an event; and thence to define the point-contact at their boundaries possible for a pair of events of which one is part of the other. We can then conceive all the intricacies of tangency. In particular we can conceive an abstractive set of which all the members have point-contact at the same event-particle. It is then easy to prove that there will be no abstractive set with the property of being covered by every abstractive set which it covers. I state this difficulty at some length because its existence guides the development of our line of argument. We have got to annex some condition to the root property of being covered by any abstractive set which it covers. When we look into this question of suitable conditions we find that in addition to event-particles all the other relevant spatial and spatio-temporal abstractive elements can be defined in the same way by suitably varying the conditions. Accordingly we proceed in a general way suitable for employment beyond event-particles.

Let σ be the name of any condition which some abstractive sets fulfil. I say that an abstractive set is

[1] Cf. 'La Théorie Relationniste de l'Espace,' *Rev. de Métaphysique et de Morale*, vol. XXIII, 1916.

'σ-prime' when it has the two properties, (i) that it satisfies the condition σ and (ii) that it is covered by every abstractive set which both is covered by it and satisfies the condition σ.

In other words you cannot get any abstractive set satisfying the condition σ which exhibits intrinsic character more simple than that of a σ-prime.

There are also the correlative abstractive sets which I call the sets of σ-antiprimes. An abstractive set is a σ-antiprime when it has the two properties, (i) that it satisfies the condition σ and (ii) that it covers every abstractive set which both covers it and satisfies the condition σ. In other words you cannot get any abstractive set satisfying the condition σ which exhibits an intrinsic character more complex than that of a σ-antiprime.

The intrinsic character of a σ-prime has a certain minimum of fullness among those abstractive sets which are subject to the condition of satisfying σ; whereas the intrinsic character of a σ-antiprime has a corresponding maximum of fullness, and includes all it can in the circumstances.

Let us first consider what help the notion of antiprimes could give us in the definition of moments which we gave in the last lecture. Let the condition σ be the property of being a class whose members are all durations. An abstractive set which satisfies this condition is thus an abstractive set composed wholly of durations. It is convenient then to define a moment as the group of abstractive sets which are equal to some σ-antiprime, where the condition σ has this special meaning. It will be found on consideration (i) that each abstractive set forming a moment is a σ-antiprime, where σ has this special meaning, and (ii) that we have excluded from membership of moments abstractive sets of durations which all have one common boundary, either the initial boundary or the final boundary. We thus exclude special cases which are apt to confuse general reasoning. The new definition of a moment, which supersedes our previous definition, is (by the aid of the notion of antiprimes) the more precisely drawn of the two, and the more useful.

The particular condition which 'σ' stood for in the definition of moments included something additional to anything which can be derived from the bare notion of extension. A duration exhibits for thought a totality. The notion of totality is something beyond that of extension, though the two are interwoven in the notion of a duration.

In the same way the particular condition 'σ' required for the definition of an event-particle must be looked for beyond the mere notion of extension. The same remark is also true of the particular conditions requisite for the other spatial elements. This additional notion is obtained by distinguishing between the notion of 'position' and the notion of convergence to an ideal zero of extension as exhibited by an abstractive set of events.

In order to understand this distinction consider a point of the instantaneous space which we conceive as apparent to us in an almost instantaneous glance. This point is an event-particle. It has two aspects. In one aspect it is there, where it is. This is its position in the space. In another aspect it is got at by ignoring the circumambient space, and by concentrating attention on the smaller and smaller set of events which approximate to it. This is its extrinsic character. Thus a point has three characters, namely, its position in the whole instantaneous space, its extrinsic character, and its intrinsic character. The same is true of any other spatial element. For example an instantaneous volume in instantaneous space has three characters, namely, its position, its extrinsic character as a group of abstractive sets, and its intrinsic character which is the limit of natural properties which is indicated by any one of these abstractive sets.

Before we can talk about position in instantaneous space, we must evidently be quite clear as to what we mean by instantaneous space in itself. Instantaneous space must be looked for as a character of a moment. For a moment is all nature at an instant. It cannot be the intrinsic character of the moment. For the intrinsic character tells us the limiting character of nature in space at that instant. Instantaneous space must be an assemblage of abstractive elements considered in their mutual relations. Thus an instantaneous space is the assemblage of abstractive elements covered by some one moment, and it is the instantaneous space of that moment.

We have now to ask what character we have found in nature which is capable of according to the elements of an instantaneous space different qualities of position. This question at once brings us to the intersection of moments, which is a topic not as yet considered in these lectures.

The locus of intersection of two moments is the assemblage of abstractive elements covered by both of them. Now two moments of the same temporal series cannot intersect. Two moments respectively of different families necessarily intersect. Accordingly in the instantaneous space of a moment we should expect the fundamental properties to be marked by the intersections with moments of other families. If M be a given moment, the intersection of M with another moment A is an instantaneous plane in the instantaneous space of M; and if B be a third moment intersecting both M and A, the intersection of M and B is another plane in the space M. Also the common intersection of A, B, and M is the intersection of the two planes in the space M, namely it is a straight line in the space M. An exceptional case arises if B and M intersect in the same plane as A and M. Furthermore if C be a fourth moment, then apart from special cases which we need not consider, it intersects M in a plane which the straight line (A, B, M) meets. Thus there is in general a common intersection of four moments of different families. This common intersection is an assemblage of abstractive elements which are each covered (or 'lie in') all four moments. The three-dimensional property of instantaneous space comes to this, that (apart from special relations between the four moments) any fifth moment either contains the whole of their common intersection or none of it. No further subdivision of the common intersection is possible by means of moments. The 'all or none' principle holds. This is not an *à priori* truth but an empirical fact of nature.

It will be convenient to reserve the ordinary spatial terms 'plane,' 'straight line,' 'point' for the elements of the timeless space of a time-system. Accordingly an instantaneous plane in the instantaneous space of a moment will be called a 'level,' an instantaneous straight line will be called a 'rect,' and an instantaneous point will be called a 'punct.' Thus a punct is the assemblage of abstractive elements which lie in each of four moments whose families have no special relations to each other. Also if P be any moment, either every abstractive element belonging to a given punct lies in P, or no abstractive element of that punct lies in P.

Position is the quality which an abstractive element possesses in virtue of the moments in which it lies. The abstractive elements which lie in the instantaneous space of a given moment M are differentiated from each other by the various other moments which intersect M so as to contain various selections of these abstractive elements. It is this differentiation of the elements which constitutes their differentiation of position. An abstractive element which belongs to a punct has the simplest type of position in M, an abstractive element which belongs to a rect but not to a punct has a more complex quality of position, an abstractive element which belongs to a level and not to a rect has a still more complex quality of position, and finally the most complex quality of position belongs to an abstractive element which belongs to a volume and not to a level. A volume however has not yet been defined. This definition will be given in the next lecture.

Evidently levels, rects, and puncts in their capacity as infinite aggregates cannot be the termini of sense-awareness, nor can they be limits which are approximated to in sense-awareness. Any one member of a level has a certain quality arising from its character as also belonging to a certain set of moments, but the level as a whole is a mere logical notion without any route of approximation along entities posited in sense-awareness.

On the other hand an event-particle is defined so as

to exhibit this character of being a route of approximation marked out by entities posited in sense-awareness. A definite event-particle is defined in reference to a definite punct in the following manner: Let the condition σ mean the property of covering all the abstractive elements which are members of that punct; so that an abstractive set which satisfies the condition σ is an abstractive set which covers every abstractive element belonging to the punct. Then the definition of the event-particle associated with the punct is that it is the group of all the σ-primes, where σ has this particular meaning.

It is evident that—with this meaning of σ—every abstractive set equal to a σ-prime is itself a σ-prime. Accordingly an event-particle as thus defined is an abstractive element, namely it is the group of those abstractive sets which are each equal to some given abstractive set. If we write out the definition of the event-particle associated with some given punct, which we will call π, it is as follows: The event-particle associated with π is the group of abstractive classes each of which has the two properties (i) that it covers every abstractive set in π and (ii) that all the abstractive sets which also satisfy the former condition as to π and which it covers, also cover it.

An event-particle has position by reason of its association with a punct, and conversely the punct gains its derived character as a route of approximation from its association with the event-particle. These two characters of a point are always recurring in any treatment of the derivation of a point from the observed facts of nature, but in general there is no clear recognition of their distinction.

The peculiar simplicity of an instantaneous point has a twofold origin, one connected with position, that is to say with its character as a punct, and the other connected with its character as an event-particle. The simplicity of the punct arises from its indivisibility by a moment.

The simplicity of an event-particle arises from the indivisibility of its intrinsic character. The intrinsic character of an event-particle is indivisible in the sense that every abstractive set covered by it exhibits the same intrinsic character. It follows that, though there are diverse abstractive elements covered by event-particles, there is no advantage to be gained by considering them since we gain no additional simplicity in the expression of natural properties.

These two characters of simplicity enjoyed respectively by event-particles and puncts define a meaning for Euclid's phrase, 'without parts and without magnitude.'

It is obviously convenient to sweep away out of our thoughts all these stray abstractive sets which are covered by event-particles without themselves being members of them. They give us nothing new in the way of intrinsic character. Accordingly we can think of rects and levels as merely loci of event-particles. In so doing we are also cutting out those abstractive elements which cover sets of event-particles, without these elements being event-particles themselves. There are classes of these abstractive elements which are of great importance. I will consider them later on in this and in other lectures. Meanwhile we will ignore them. Also I will always speak of 'event-particles' in preference to 'puncts,' the latter being an artificial word for which I have no great affection.

Parallelism among rects and levels is now explicable.

Consider the instantaneous space belonging to a moment *A*, and let *A* belong to the temporal series of moments which I will call α. Consider any other temporal series of moments which I will call β. The moments of β do not intersect each other and they intersect the moment *A* in a family of levels. None of these levels can intersect, and they form a family of parallel instantaneous planes in the instantaneous space of moment *A*. Thus the parallelism of moments in a temporal series begets the parallelism of levels in an instantaneous space, and thence—as it is easy to see—the parallelism of rects. Accordingly the Euclidean property of space arises from the parabolic property of time. It may be that there is reason to adopt a hyperbolic theory of time and a corresponding hyperbolic theory of space. Such a theory has not been worked out, so it is not possible to judge as to the character of the evidence which could be brought forward in its favour.

The theory of order in an instantaneous space is immediately derived from time-order. For consider the space of a moment *M*. Let α be the name of a time-system to which *M* does not belong. Let A_1, A_2, A_3, etc. be moments of α in the order of their occurrences. Then A_1, A_2, A_3, etc. intersect *M* in parallel levels l_1, l_2, l_3, etc. Then the relative order of the parallel levels in the space of *M* is the same as the relative order of the corresponding moments in the time-system α. Any rect in *M* which intersects all these levels in its set of puncts, thereby receives for its puncts an order of position on it. So spatial order is derivative from temporal order. Furthermore there are alternative time-systems, but there is only one definite spatial order in each instantaneous space. Accordingly the various modes of deriving spatial order from diverse time-systems must harmonise with one spatial order in each instantaneous space. In this way also diverse time-orders are comparable.

We have two great questions still on hand to be settled before our theory of space is fully adjusted. One of these is the question of the determination of the methods of measurement within the space, in other words, the congruence-theory of the space. The measurement of space will be found to be closely connected with the measurement of time, with respect to which no principles have as yet been determined. Thus our congruence-theory will be a theory both for space and for time. Secondly there is the determination of the timeless space which corresponds to any particular time-system with its infinite set of instantaneous spaces in its successive moments. This is the space—or rather, these are the spaces—of physical science. It is very usual to dismiss this space by saying that this is conceptual. I do not understand the virtue of these phrases. I suppose that it is meant that the space is the conception of something in nature. Accordingly if the space of physical science is to be called conceptual, I ask, What in nature is it the conception of? For example, when we speak of a point in the timeless space of physical science, I suppose that we are speaking of something in nature. If we are not so speaking, our scientists are exercising their wits in the realms of pure fantasy, and this is palpably not the case. This demand for a definite Habeas Corpus Act for the production of the relevant entities in nature applies whether space be relative or absolute. On the theory of relative space, it may perhaps be argued that there is no timeless space for physical science, and that there is only the momentary series of instantaneous spaces.

An explanation must then be asked for the meaning of the very common statement that such and such a man walked four miles in some definite hour. How can you measure distance from one space into another space? I understand walking out of the sheet of an ordnance map. But the meaning of saying that Cambridge at 10 o'clock this morning in the appropriate instantaneous space for that instant is 52 miles from London at 11 o'clock this morning in the appropriate instantaneous space for that instant beats me entirely. I think that, by the time a meaning has been produced for this statement, you will find that you have constructed what is in fact a timeless space. What I cannot understand is how to produce an explanation of meaning without in effect making some such construction. Also I may add that I do not know how the instantaneous spaces are thus correlated into one space by any method which is available on the current theories of space.

You will have noticed that by the aid of the assumption of alternative time-systems, we are arriving at an explanation of the character of space. In natural science 'to explain' means merely to discover 'interconnexions.' For example, in one sense there is no explanation of the red which you see. It is red, and there is nothing else to be said about it. Either it is posited before you in sense-awareness or you are ignorant of the entity red. But science has explained red. Namely it has discovered interconnexions between red as a factor in nature and other factors in nature, for example waves of light which are waves of electromagnetic disturbances.

There are also various pathological states of the body which lead to the seeing of red without the occurrence of light waves. Thus connexions have been discovered between red as posited in sense-awareness and various other factors in nature. The discovery of these connexions constitutes the scientific explanation of our vision of colour. In like manner the dependence of the character of space on the character of time constitutes an explanation in the sense in which science seeks to explain. The systematising intellect abhors bare facts. The character of space has hitherto been presented as a collection of bare facts, ultimate and disconnected. The theory which I am expounding sweeps away this disconnexion of the facts of space.

CHAPTER V

SPACE AND MOTION

THE topic for this lecture is the continuation of the task of explaining the construction of spaces as abstracts from the facts of nature. It was noted at the close of the previous lecture that the question of congruence had not been considered, nor had the construction of a timeless space which should correlate the successive momentary spaces of a given time-system. Furthermore it was also noted that there were many spatial abstractive elements which we had not yet defined. We will first consider the definition of some of these abstractive elements, namely the definitions of solids, of areas, and of routes. By a 'route' I mean a linear segment, whether straight or curved. The exposition of these definitions and the preliminary explanations necessary will, I hope, serve as a general explanation of the function of event-particles in the analysis of nature.

We note that event-particles have 'position' in respect to each other. In the last lecture I explained that 'position' was quality gained by a spatial element in virtue of the intersecting moments which covered it. Thus each event-particle has position in this sense. The simplest mode of expressing the position in nature of an event-particle is by first fixing on any definite time-system. Call it α. There will be one moment of the temporal series of α which covers the given event-particle. Thus the position of the event-particle in the temporal series α is defined by this moment, which we will call M. The position of the particle in the space of M is then fixed in the ordinary way by three levels which intersect in it and in it only. This procedure of fixing the position of an event-particle shows that the aggregate of event-particles forms a four-dimensional manifold. A finite event occupies a limited chunk of this manifold in a sense which I now proceed to explain.

Let *e* be any given event. The manifold of event-particles falls into three sets in reference to *e*. Each event-particle is a group of equal abstractive sets and each abstractive set towards its small-end is composed of smaller and smaller finite events. When we select from these finite events which enter into the make-up of a given event-particle those which are small enough, one of three cases must occur. Either (i) all of these small events are entirely separate from the given event *e*, or (ii) all of these small events are parts of the event *e*, or (iii) all of these small events overlap the event *e* but are not parts of it. In the first case the event-particle will be said to 'lie outside' the event *e*, in the second case the event-particle will be said to 'lie inside' the event *e*, and in the third case the event-particle will be said to be a 'boundary-particle' of the event *e*. Thus there are three sets of particles, namely the set of those which lie outside the event *e*, the set of those which lie inside the event *e*, and the boundary of the event *e* which is the set of boundary-particles of *e*. Since an event is four-dimensional, the boundary of an event is a three-dimensional manifold. For a finite event there is a continuity of boundary; for a duration the boundary consists of those event-particles which are covered by either of the two bounding moments. Thus the boundary of a duration consists of two momentary three-dimensional spaces. An event will be said to 'occupy' the aggregate of event-particles which lie within it.

Two events which have 'junction' in the sense in which junction was described in my last lecture, and yet are separated so that neither event either overlaps or is part of the other event, are said to be 'adjoined.'

This relation of adjunction issues in a peculiar relation between the boundaries of the two events. The two boundaries must have a common portion which is in fact a continuous three-dimensional locus of event-particles in the four-dimensional manifold.

A three-dimensional locus of event-particles which is the common portion of the boundary of two adjoined events will be called a 'solid.' A solid may or may not lie completely in one moment. A solid which does not lie in one moment will be called 'vagrant.' A solid which does lie in one moment will be called a volume. A volume may be defined as the locus of the event-particles in which a moment intersects an event, provided that the two do intersect. The intersection of a moment and an event will evidently consist of those event-particles which are covered by the moment and lie in the event. The identity of the two definitions of a volume is evident when we remember that an intersecting moment divides the event into two adjoined events.

A solid as thus defined, whether it be vagrant or be a volume, is a mere aggregate of event-particles illustrating a certain quality of position. We can also define a solid as an abstractive element. In order to do so we recur to the theory of primes explained in the preceding lecture. Let the condition named σ stand for the fact that each of the events of any abstractive set satisfying it has all the event-particles of some particular solid lying in it. Then the group of all the σ-primes is the abstractive element which is associated with the given solid. I will call this abstractive element the solid as an abstractive element, and I will call the aggregate of event-particles the solid as a locus. The instantaneous volumes in instantaneous space which are the ideals of our sense-perception are volumes as abstractive elements. What we really perceive with all our efforts after exactness are small events far enough down some abstractive set belonging to the volume as an abstractive element.

It is difficult to know how far we approximate to any perception of vagrant solids. We certainly do not think that we make any such approximation. But then our thoughts—in the case of people who do think about such topics—are so much under the control of the materialistic theory of nature that they hardly count for evidence. If Einstein's theory of gravitation has any truth in it, vagrant solids are of great importance in science. The whole boundary of a finite event may be looked on as a particular example of a vagrant solid as a locus. Its particular property of being closed prevents it from being definable as an abstractive element.

When a moment intersects an event, it also intersects the boundary of that event. This locus, which is the portion of the boundary contained in the moment, is the bounding surface of the corresponding volume of that event contained in the moment. It is a two-dimensional locus.

The fact that every volume has a bounding surface is the origin of the Dedekindian continuity of space.

Another event may be cut by the same moment in another volume and this volume will also have its boundary. These two volumes in the instantaneous space of one moment may mutually overlap in the familiar way which I need not describe in detail and thus cut off portions from each other's surfaces. These portions of surfaces are 'momental areas.'

It is unnecessary at this stage to enter into the complexity of a definition of vagrant areas. Their definition is simple enough when the four-dimensional manifold of event-particles has been more fully explored as to its properties.

Momental areas can evidently be defined as abstractive elements by exactly the same method as applied to solids. We have merely to substitute 'area' for a 'solid' in the words of the definition already given. Also, exactly as in the analogous case of a solid, what we perceive as an approximation to our ideal of an area is a small event far enough down towards the small end of one of the equal abstractive sets which belongs to the area as an abstractive element.

Two momental areas lying in the same moment can cut each other in a momental segment which is not necessarily rectilinear. Such a segment can also be defined as an abstractive element. It is then called a 'momental route.' We will not delay over any general consideration of these momental routes, nor is it important for us to proceed to the still wider investigation of vagrant routes in general. There are however two simple sets of routes which are of vital importance. One is a set of momental routes and the other of vagrant routes. Both sets can be classed together as straight routes. We proceed to define them without any reference to the definitions of volumes and surfaces.

The two types of straight routes will be called rectilinear routes and stations. Rectilinear routes are momental routes and stations are vagrant routes. Rectilinear routes are routes which in a sense lie in rects. Any two event-particles on a rect define the set of event-particles which lie between them on that rect. Let the satisfaction of the condition σ by an abstractive set mean that the two given event-particles and the event-particles lying between them on the rect all lie in every event belonging to the abstractive set. The group of σ-primes, where σ has this meaning, form an abstractive element. Such abstractive elements are rectilinear routes. They are the segments of instantaneous straight lines which are the ideals of exact perception. Our actual perception, however exact, will be the perception of a small event sufficiently far down one of the abstractive sets of the abstractive element.

A station is a vagrant route and no moment can intersect any station in more than one event-particle. Thus a station carries with it a comparison of the positions in their respective moments of the event-particles covered by it. Rects arise from the intersection of moments. But as yet no properties of events have been mentioned by which any analogous vagrant loci can be found out.

The general problem for our investigation is to determine a method of comparison of position in one instantaneous space with positions in other instantaneous spaces. We may limit ourselves to the spaces of the parallel moments of one time-system. How are positions in these various spaces to be compared? In other words, What do we mean by motion? It is the fundamental question to be asked of any theory of relative space, and like many other fundamental questions it is apt to be left unanswered. It is not an answer to reply, that we all know what we mean by motion. Of course we do, so far as sense-awareness is concerned. I am asking that your theory of space should provide nature with something to be observed. You have not settled the question by bringing forward a theory according to which there is nothing to be observed, and by then reiterating that nevertheless we do observe this non-existent fact. Unless motion is something as a fact in nature, kinetic energy and momentum and all that depends on these physical concepts evaporate from our list of physical realities. Even in this revolutionary age my conservatism resolutely opposes the identification of momentum and moonshine.

Accordingly I assume it as an axiom, that motion is a physical fact. It is something that we perceive as in nature. Motion presupposes rest. Until theory arose to vitiate immediate intuition, that is to say to vitiate the uncriticised judgments which immediately arise from sense-awareness, no one doubted that in motion you leave behind that which is at rest. Abraham in his wanderings left his birthplace where it had ever been. A theory of motion and a theory of rest are the same thing viewed from different aspects with altered emphasis.

Now you cannot have a theory of rest without in some sense admitting a theory of absolute position. It is usually assumed that relative space implies that there is no absolute position. This is, according to my creed, a mistake. The assumption arises from the failure to make another distinction; namely, that there may be alternative definitions of absolute position. This possibility enters with the admission of alternative time-systems. Thus the series of spaces in the parallel moments of one temporal series may have their own definition of absolute position correlating sets of event-particles in these successive spaces, so that each set consists of event-particles, one from each space, all with the property of possessing the same absolute position in that series of spaces. Such a set of event-particles will form a point in the timeless space of that time-system. Thus a point is really an absolute position in the timeless space of a given time-system.

But there are alternative time-systems, and each time-system has its own peculiar group of points—that is to say, its own peculiar definition of absolute position. This is exactly the theory which I will elaborate.

In looking to nature for evidence of absolute position it is of no use to recur to the four-dimensional manifold of event-particles. This manifold has been obtained by the extension of thought beyond the immediacy of observation. We shall find nothing in it except what we have put there to represent the ideas in thought which arise from our direct sense-awareness of nature. To find evidence of the properties which are to be found in the manifold of event-particles we must always recur to the observation of relations between events. Our problem is to determine those relations between events which issue in the property of absolute position in a timeless space. This is in fact the problem of the determination of the very meaning of the timeless spaces of physical science.

In reviewing the factors of nature as immediately disclosed in sense-awareness, we should note the fundamental character of the percept of 'being here.' We discern an event merely as a factor in a determinate complex in which each factor has its own peculiar share.

There are two factors which are always ingredient in this complex, one is the duration which is represented in thought by the concept of all nature that is present now, and the other is the peculiar *locus standi* for mind involved in the sense-awareness. This *locus standi* in nature is what is represented in thought by the concept of 'here,' namely of an 'event here.'

This is the concept of a definite factor in nature. This factor is an event in nature which is the focus in nature for that act of awareness, and the other events are perceived as referred to it. This event is part of the associated duration. I call it the 'percipient event.' This event is not the mind, that is to say, not the percipient. It is that in nature from which the mind perceives. The complete foothold of the mind in nature is represented by the pair of events, namely, the present duration which marks the 'when' of awareness and the percipient event which marks the 'where' of awareness and the 'how' of awareness. This percipient event is roughly speaking the bodily life of the incarnate mind. But this identification is only a rough one. For the functions of the body shade off into those of other events in nature; so that for some purposes the percipient event is to be reckoned as merely part of the bodily life and for other purposes it may even be reckoned as more than the bodily life. In many respects the demarcation is purely arbitrary, depending upon where in a sliding scale you choose to draw the line.

I have already in my previous lecture on Time discussed the association of mind with nature. The difficulty of the discussion lies in the liability of constant factors to be overlooked. We never note them by contrast with their absences. The purpose of a discussion of such factors may be described as being to make obvious things look odd. We cannot envisage them unless we manage to invest them with some of the freshness which is due to strangeness.

It is because of this habit of letting constant factors slip from consciousness that we constantly fall into the error of thinking of the sense-awareness of a particular factor in nature as being a two-termed relation between the mind and the factor. For example, I perceive a green leaf. Language in this statement suppresses all reference to any factors other than the percipient mind and the green leaf and the relation of sense-awareness. It discards the obvious inevitable factors which are essential elements in the perception. I am here, the leaf is there; and the event here and the event which is the life of the leaf there are both embedded in a totality of nature which is now, and within this totality there are other discriminated factors which it is irrelevant to mention. Thus language habitually sets before the mind a misleading abstract of the indefinite complexity of the fact of sense-awareness.

What I now want to discuss is the special relation of the percipient event which is 'here' to the duration which is 'now.' This relation is a fact in nature, namely the mind is aware of nature as being with these two factors in this relation.

Within the short present duration the 'here' of the percipient event has a definite meaning of some sort. This meaning of 'here' is the content of the special relation of the percipient event to its associated duration. I will call this relation 'cogredience.' Accordingly I ask for a description of the character of the relation of cogredience. The present snaps into a past and a present when the 'here' of cogredience loses its single determinate meaning. There has been a passage of nature from the 'here' of perception within the past duration to the different 'here' of perception within the present duration. But the two 'heres' of sense-awareness within neighbouring durations may be indistinguishable. In this case there has been a passage from the past to the present, but a more retentive perceptive force might have retained the passing nature as one complete present instead of letting the earlier duration slip into the past. Namely, the sense of rest helps the integration of durations into a prolonged present, and the sense of motion differentiates nature into a succession of shortened durations. As we look out of a railway carriage in an express train, the present is past before reflexion can seize it. We live in snippits too quick for thought. On the other hand the immediate present is prolonged according as nature presents itself to us in an aspect of unbroken rest. Any change in nature provides ground for a differentiation among durations so as to shorten the present. But there is a great distinction between self-change in nature and change in external nature. Self-change in nature is change in the quality of the standpoint of the percipient event. It is the break up of the 'here' which necessitates the break up of the present duration. Change in external nature is compatible with a prolongation of the present of contemplation rooted in a given standpoint. What I want to bring out is that the preservation of a peculiar relation to a duration is a necessary condition for the function of that duration as a present duration for sense-awareness. This peculiar relation is the relation of cogredience between the percipient event and the duration.

Cogredience is the preservation of unbroken quality of standpoint within the duration. It is the continuance of identity of station within the whole of nature which is the terminus of sense-awareness. The duration may comprise change within itself, but cannot—so far as it is one present duration—comprise change in the quality of its peculiar relation to the contained percipient event.

In other words, perception is always 'here,' and a duration can only be posited as present for sense-awareness on condition that it affords one unbroken meaning of 'here' in its relation to the percipient event. It is only in the past that you can have been 'there' with a standpoint distinct from your present 'here.'

Events there and events here are facts of nature, and the qualities of being 'there' and 'here' are not merely qualities of awareness as a relation between nature and mind. The quality of determinate station in the duration which belongs to an event which is 'here' in one determinate sense of 'here' is the same kind of quality of station which belongs to an event which is 'there' in one determinate sense of 'there.' Thus cogredience has nothing to do with any biological character of the event which is related by it to the associated duration. This biological character is apparently a further condition for the peculiar connexion of a percipient event with the percipience of mind; but it has nothing to do with the relation of the percipient event to the duration which is the present whole of nature posited as the disclosure of the percipience.

Given the requisite biological character, the event in its character of a percipient event selects that duration with which the operative past of the event is practically cogredient within the limits of the exactitude of observation. Namely, amid the alternative time-systems which nature offers there will be one with a duration giving the best average of cogredience for all the subordinate parts of the percipient event. This duration will be the whole of nature which is the terminus posited by sense-awareness. Thus the character of the percipient event determines the time-system immediately evident in nature. As the character of the percipient event changes with the passage of nature—or, in other words, as the percipient mind in its passage correlates itself with the passage of the percipient event into another percipient event—the time-system correlated with the percipience of that mind may change. When the bulk of the events perceived are cogredient in a duration other than that of the percipient event, the percipience may include a double consciousness of cogredience, namely the consciousness of the whole within which the observer in the train is 'here,' and the consciousness of the whole within which the trees and bridges and telegraph posts are definitely 'there.' Thus in perceptions under certain circumstances the events discriminated assert their own relations of cogredience. This assertion of cogredience is peculiarly evident when the duration to which the perceived event is cogredient is the same as the duration which is the present whole of nature—in other words, when the event and the percipient event are both cogredient to the same duration.

We are now prepared to consider the meaning of stations in a duration, where stations are a peculiar kind of routes, which define absolute position in the associated timeless space.

There are however some preliminary explanations. A finite event will be said to extend throughout a duration when it is part of the duration and is intersected by any moment which lies in the duration. Such an event begins with the duration and ends with it. Furthermore every event which begins with the duration and ends with it, extends throughout the duration. This is an axiom based on the continuity of events. By beginning with a duration and ending with it, I mean (i) that the event is part of the duration, and (ii) that both the initial and final boundary moments of the duration cover some event-particles on the boundary of the event.

Every event which is cogredient with a duration extends throughout that duration.

It is not true that all the parts of an event cogredient with a duration are also cogredient with the duration. The relation of cogredience may fail in either of two ways. One reason for failure may be that the part does not extend throughout the duration. In this case the part may be cogredient with another duration which is part of the given duration, though it is not cogredient with the given duration itself. Such a part would be cogredient if its existence were sufficiently prolonged in that time-system. The other reason for failure arises from the four-dimensional extension of events so that there is no determinate route of transition of events in linear series. For example, the tunnel of a tube railway is an event at rest in a certain time-system, that is to say, it is cogredient with a certain duration. A train travelling in it is part of that tunnel, but is not itself at rest.

If an event e be cogredient with a duration d, and d' be any duration which is part of d. Then d' belongs to the same time-system as d. Also d' intersects e in an event e' which is part of e and is cogredient with d'.

Let P be any event-particle lying in a given duration d. Consider the aggregate of events in which P lies and which are also cogredient with d. Each of these events occupies its own aggregate of event-particles. These aggregates will have a common portion, namely the class of event-particle lying in all of them. This class of event-particles is what I call the 'station' of the event-particle P in the duration d. This is the station in the character of a locus. A station can also be defined in the character of an abstractive element. Let the property σ be the name of the property which an abstractive set possesses when (i) each of its events is cogredient with the duration d and (ii) the event-particle P lies in each of its events. Then the group of σ-primes, where σ has this meaning, is an abstractive element and is the station of P in d as an abstractive element. The locus of event-particles covered by the station of P in d as an abstractive element is the station of P in d as a locus. A station has accordingly the usual three characters, namely, its character of position, its extrinsic character as an abstractive element, and its intrinsic character.

It follows from the peculiar properties of rest that two stations belonging to the same duration cannot intersect. Accordingly every event-particle on a station of a duration has that station as its station in the duration. Also every duration which is part of a given duration intersects the stations of the given duration in loci which are its own stations. By means of these properties we can utilise the overlappings of the durations of one family—that is, of one time-system—to prolong stations indefinitely backwards and forwards. Such a prolonged station will be called a point-track. A point-track is a locus of event-particles. It is defined by reference to one particular time-system, α say. Corresponding to any other time-system these will be a different group of point-tracks. Every event-particle will lie on one and only one point-track of the group belonging to any one time-system. The group of point-tracks of the time-system α is the group of points of the timeless space of α. Each such point indicates a certain quality of absolute position in reference to the durations of the family associated with α, and thence in reference to the successive instantaneous spaces lying in the successive moments of α. Each moment of α will intersect a point-track in one and only one event-particle.

This property of the unique intersection of a moment and a point-track is not confined to the case when the moment and the point-track belong to the same time-system. Any two event-particles on a point-track are sequential, so that they cannot lie in the same moment. Accordingly no moment can intersect a point-track more than once, and every moment intersects a point-track in one event-particle.

Anyone who at the successive moments of α should be at the event-particles where those moments intersect a given point of α will be at rest in the timeless space of time-system α. But in any other timeless space belonging to another time-system he will be at a different point at each succeeding moment of that time-system. In other words he will be moving. He will be moving in a straight line with uniform velocity. We might take this as the definition of a straight line. Namely, a straight line in the space of time-system β is the locus of those points of β which all intersect some one point-track which is a point in the space of some other time-system. Thus each point in the space of a time-system α is associated with one and only one straight line of the space of any other time-system β. Furthermore the set of straight lines in space β which are thus associated with points in space α form a complete family of parallel straight lines in space β. Thus there is a one-to-one correlation of points in space α with the straight lines of a certain definite family of parallel straight lines in space β. Conversely there is an analogous one-to-one correlation of the points in space β with the straight lines of a certain family of parallel straight lines in space α. These families will be called respectively the family of parallels in β associated with α, and the family of parallels in α associated with β. The direction in the space of β indicated by the family of parallels in β will be called the direction of α in space β, and the family of parallels in α is the direction of β in space α. Thus a being at rest at a point of space α will be moving uniformly along a line in space β which is in the direction of α in space β, and a being at rest at a point of space β will be moving uniformly along a line in space α which is in the direction of β in space α.

I have been speaking of the timeless spaces which are associated with time-systems. These are the spaces of physical science and of any concept of space as eternal and unchanging. But what we actually perceive is an approximation to the instantaneous space indicated by event-particles which lie within some moment of the time-system associated with our awareness. The points of such an instantaneous space are event-particles and the straight lines are rects. Let the time-system be named α, and let the moment of time-system α to which our quick perception of nature approximates be called *M*. Any straight line *r* in space α is a locus of points and each point is a point-track which is a locus of event-particles. Thus in the four-dimensional geometry of all event-particles there is a two-dimensional locus which is the locus of all event-particles on points lying on the straight line *r*. I will call this locus of event-particles the matrix of the straight line *r*. A matrix intersects any moment in a rect. Thus the matrix of *r* intersects the moment *M* in a rect ρ. Thus ρ is the instantaneous rect in *M* which occupies at the moment *M* the straight line *r* in the space of α. Accordingly when one sees instantaneously a moving being and its path ahead of it, what one really sees is the being at some event-particle *A* lying in the rect ρ which is the apparent path on the assumption of uniform motion. But the actual rect ρ which is a locus of event-particles is never traversed by the being. These event-particles are the instantaneous facts which pass with the instantaneous moment. What is really traversed are other event-particles which at succeeding instants occupy the same points of space α as those occupied by the event-particles of the rect ρ. For example, we see a stretch of road and a lorry moving along it. The instantaneously seen road is a portion of the rect ρ—of course only an approximation to it. The lorry is the moving object. But the road as seen is never traversed. It is thought of as being traversed because the intrinsic characters of the later events are in general so similar to those of the instantaneous road that we do not trouble to discriminate. But suppose a land mine under the road has been exploded before the lorry gets there. Then it is fairly obvious that the lorry does not traverse what we saw at first. Suppose the lorry is at rest in space β. Then the straight line r of space α is in the direction of β in space α, and the rect ρ is the representative in the moment M of the line r of space α. The direction of ρ in the instantaneous space of the moment M is the direction of β in M, where M is a moment of time-system α. Again the matrix of the line r of space α will also be the matrix of some line s of space β which will be in the direction of α in space β. Thus if the lorry halts at some point P of space α which lies on the line r, it is now moving along the line s of space β. This is the theory of relative motion; the common matrix is the bond which connects the motion of β in space α with the motions of α in space β.

Motion is essentially a relation between some object of nature and the one timeless space of a time-system. An instantaneous space is static, being related to the static nature at an instant. In perception when we see things moving in an approximation to an instantaneous space, the future lines of motion as immediately perceived are rects which are never traversed. These approximate rects are composed of small events, namely approximate routes and event-particles, which are passed away before the moving objects reach them. Assuming that our forecasts of rectilinear motion are correct, these rects occupy the straight lines in timeless space which are traversed. Thus the rects are symbols in immediate sense-awareness of a future which can only be expressed in terms of timeless space.

We are now in a position to explore the fundamental character of perpendicularity. Consider the two time-systems α and β, each with its own timeless space and its own family of instantaneous moments with their instantaneous spaces. Let M and N be respectively a moment of α and a moment of β. In M there is the direction of β and in N there is the direction of α. But M and N, being moments of different time-systems, intersect in a level. Call this level λ. Then λ is an instantaneous plane in the instantaneous space of M and also in the instantaneous space of N. It is the locus of all the event-particles which lie both in M and in N.

In the instantaneous space of M the level λ is perpendicular to the direction of β in M, and in the instantaneous space of N the level λ is perpendicular to the direction of α in N. This is the fundamental property which forms the definition of perpendicularity. The symmetry of perpendicularity is a particular instance of the symmetry of the mutual relations between two time-systems. We shall find in the next lecture that it is from this symmetry that the theory of congruence is deduced.

The theory of perpendicularity in the timeless space of any time-system α follows immediately from this theory of perpendicularity in each of its instantaneous spaces. Let ρ be any rect in the moment M of α and let λ be a level in M which is perpendicular to ρ. The locus of those points of the space of α which intersect M in event-particles on ρ is the straight line r of space α, and the locus of those points of the space of α which intersect M in event-particles on λ is the plane l of space α. Then the plane l is perpendicular to the line r.

In this way we have pointed out unique and definite properties in nature which correspond to perpendicularity. We shall find that this discovery of definite unique properties defining perpendicularity is of critical importance in the theory of congruence which is the topic for the next lecture.

I regret that it has been necessary for me in this lecture to administer such a large dose of four-dimensional geometry. I do not apologise, because I am really not responsible for the fact that nature in its most fundamental aspect is four-dimensional. Things are what they are; and it is useless to disguise the fact that 'what things are' is often very difficult for our intellects to follow. It is a mere evasion of the ultimate problems to shirk such obstacles.

CHAPTER VI

CONGRUENCE

THE aim of this lecture is to establish a theory of congruence. You must understand at once that congruence is a controversial question. It is the theory of measurement in space and in time. The question seems simple. In fact it is simple enough for a standard procedure to have been settled by act of parliament; and devotion to metaphysical subtleties is almost the only crime which has never been imputed to any English parliament. But the procedure is one thing and its meaning is another.

First let us fix attention on the purely mathematical question. When the segment between two points A and B is congruent to that between the two points C and D, the quantitative measurements of the two segments are equal. The equality of the numerical measures and the congruence of the two segments are not always clearly discriminated, and are lumped together under the term equality. But the procedure of measurement presupposes congruence. For example, a yard measure is applied successively to measure two distances between two pairs of points on the floor of a room. It is of the essence of the procedure of measurement that the yard measure remains unaltered as it is transferred from one position to another. Some objects can palpably alter as they move—for example, an elastic thread; but a yard measure does not alter if made of the proper material. What is this but a judgment of congruence applied to the train of successive positions of the yard measure? We know that it does not alter because we judge it to be congruent to itself in various positions. In the case of the thread we can observe the loss of self-congruence. Thus immediate judgments of congruence are presupposed in measurement, and the process of measurement is merely a procedure to extend the recognition of congruence to cases where these immediate judgments are not available. Thus we cannot define congruence by measurement.

In modern expositions of the axioms of geometry certain conditions are laid down which the relation of congruence between segments is to satisfy. It is supposed that we have a complete theory of points, straight lines, planes, and the order of points on planes—in fact, a complete theory of non-metrical geometry. We then enquire about congruence and lay down the set of conditions—or axioms as they are called—which this relation satisfies. It has then been proved that there are alternative relations which satisfy these conditions equally well and that there is nothing intrinsic in the theory of space to lead us to adopt any one of these relations in preference to any other as the relation of congruence which we adopt. In other words there are alternative metrical geometries which all exist by an equal right so far as the intrinsic theory of space is concerned.

Poincaré, the great French mathematician, held that our actual choice among these geometries is guided purely by convention, and that the effect of a change of choice would be simply to alter our expression of the physical laws of nature. By 'convention' I understand Poincaré to mean that there is nothing inherent in nature itself giving any peculiar *rôle* to one of these congruence relations, and that the choice of one particular relation is guided by the volitions of the mind at the other end of the sense-awareness. The principle of guidance is intellectual convenience and not natural fact.

This position has been misunderstood by many of Poincaré's expositors. They have muddled it up with another question, namely that owing to the inexactitude of observation it is impossible to make an exact statement in the comparison of measures. It follows that a certain subset of closely allied congruence relations can be assigned of which each member equally well agrees with that statement of observed congruence when the statement is properly qualified with its limits of error.

This is an entirely different question and it presupposes a rejection of Poincaré's position. The absolute indetermination of nature in respect of all the relations of congruence is replaced by the indetermination of observation with respect to a small subgroup of these relations.

Poincaré's position is a strong one. He in effect challenges anyone to point out any factor in nature which gives a preeminent status to the congruence relation which mankind has actually adopted. But undeniably the position is very paradoxical. Bertrand Russell had a controversy with him on this question, and pointed out that on Poincaré's principles there was nothing in nature to determine whether the earth is larger or smaller than some assigned billiard ball. Poincaré replied that the attempt to find reasons in nature for the selection of a definite congruence relation in space is like trying to determine the position of a ship in the ocean by counting the crew and observing the colour of the captain's eyes.

In my opinion both disputants were right, assuming the grounds on which the discussion was based. Russell in effect pointed out that apart from minor inexactitudes a determinate congruence relation is among the factors in nature which our sense-awareness posits for us. Poincaré asks for information as to the factor in nature which might lead any particular congruence relation to play a preeminent *rôle* among the factors posited in sense-awareness. I cannot see the answer to either of these contentions provided that you admit the materialistic theory of nature. With this theory nature at an instant in space is an independent fact. Thus we have to look for our preeminent congruence relation amid nature in instantaneous space; and Poincaré is undoubtedly right in saying that nature on this hypothesis gives us no help in finding it.

On the other hand Russell is in an equally strong position when he asserts that, as a fact of observation, we do find it, and what is more agree in finding the same congruence relation. On this basis it is one of the most extraordinary facts of human experience that all mankind without any assignable reason should agree in fixing attention on just one congruence relation amid the indefinite number of indistinguishable competitors for notice. One would have expected disagreement on this fundamental choice to have divided nations and to have rent families. But the difficulty was not even discovered till the close of the nineteenth century by a few mathematical philosophers and philosophic mathematicians. The case is not like that of our agreement on some fundamental fact of nature such as the three dimensions of space. If space has only three dimensions we should expect all mankind to be aware of the fact, as they are aware of it. But in the case of congruence, mankind agree in an arbitrary interpretation of sense-awareness when there is nothing in nature to guide it.

I look on it as no slight recommendation of the theory of nature which I am expounding to you that it gives a solution of this difficulty by pointing out the factor in nature which issues in the preeminence of one congruence relation over the indefinite herd of other such relations.

The reason for this result is that nature is no longer confined within space at an instant. Space and time are now interconnected; and this peculiar factor of time which is so immediately distinguished among the deliverances of our sense-awareness, relates itself to one particular congruence relation in space.

Congruence is a particular example of the fundamental fact of recognition. In perception we recognise. This recognition does not merely concern the comparison of a factor of nature posited by memory with a factor posited by immediate sense-awareness. Recognition takes place within the present without any intervention of pure memory. For the present fact is a duration with its antecedent and consequent durations which are parts of itself. The discrimination in sense-awareness of a finite event with its quality of passage is also accompanied by the discrimination of other factors of nature which do not share in the passage of events. Whatever passes is an event. But we find entities in nature which do not pass; namely we recognise samenesses in nature. Recognition is not primarily an intellectual act of comparison; it is in its essence merely sense-awareness in its capacity of positing before us factors in nature which do not pass. For example, green is perceived as situated in a certain finite event within the present duration. This green preserves its self-identity throughout, whereas the event passes and thereby obtains the property of breaking into parts. The green patch has parts. But in talking of the green patch we are speaking of the event in its sole capacity of being for us the situation of green. The green itself is numerically one self-identical entity, without parts because it is without passage.

Factors in nature which are without passage will be called objects. There are radically different kinds of objects which will be considered in the succeeding lecture.

Recognition is reflected into the intellect as comparison. The recognised objects of one event are compared with the recognised objects of another event. The comparison may be between two events in the present, or it may be between two events of which one is posited by memory-awareness and the other by immediate sense-awareness. But it is not the events which are compared. For each event is essentially unique and incomparable. What are compared are the objects and relations of objects situated in events. The event considered as a relation between objects has lost its passage and in this aspect is itself an object. This object is not the event but only an intellectual abstraction. The same object can be situated in many events; and in this sense even the whole event, viewed as an object, can recur, though not the very event itself with its passage and its relations to other events.

Objects which are not posited by sense-awareness may be known to the intellect. For example, relations between objects and relations between relations may be factors in nature not disclosed in sense-awareness but known by logical inference as necessarily in being. Thus objects for our knowledge may be merely logical abstractions. For example, a complete event is never disclosed in sense-awareness, and thus the object which is the sum total of objects situated in an event as thus inter-related is a mere abstract concept. Again a right-angle is a perceived object which can be situated in many events; but, though rectangularity is posited by sense-awareness, the majority of geometrical relations are not so posited. Also rectangularity is in fact often not perceived when it can be proved to have been there for perception. Thus an object is often known merely as an abstract relation not directly posited in sense-awareness although it is there in nature.

The identity of quality between congruent segments is generally of this character. In certain special cases this identity of quality can be directly perceived. But in general it is inferred by a process of measurement depending on our direct sense-awareness of selected cases and a logical inference from the transitive character of congruence.

Congruence depends on motion, and thereby is generated the connexion between spatial congruence and temporal congruence. Motion along a straight line has a symmetry round that line. This symmetry is expressed by the symmetrical geometrical relations of the line to the family of planes normal to it.

Also another symmetry in the theory of motion arises from the fact that rest in the points of β corresponds to uniform motion along a definite family of parallel straight lines in the space of α. We must note the three characteristics, (i) of the uniformity of the motion corresponding to any point of β along its correlated straight line in α, and (ii) of the equality in magnitude of the velocities along the various lines of α correlated to rest in the various points of β, and (iii) of the parallelism of the lines of this family.

We are now in possession of a theory of parallels and a theory of perpendiculars and a theory of motion, and from these theories the theory of congruence can be constructed. It will be remembered that a family of parallel levels in any moment is the family of levels in which that moment is intersected by the family of moments of some other time-system. Also a family of parallel moments is the family of moments of some one time-system. Thus we can enlarge our concept of a family of parallel levels so as to include levels in different moments of one time-system. With this enlarged concept we say that a complete family of parallel levels in a time-system α is the complete family of levels in which the moments of α intersect the moments of β. This complete family of parallel levels is also evidently a family lying in the moments of the time-system β. By introducing a third time-system γ, parallel rects are obtained. Also all the points of any one time-system form a family of parallel point-tracks. Thus there are three types of parallelograms in the four-dimensional manifold of event-particles.

In parallelograms of the first type the two pairs of parallel sides are both of them pairs of rects. In parallelograms of the second type one pair of parallel sides is a pair of rects and the other pair is a pair of point-tracks. In parallelograms of the third type the two pairs of parallel sides are both of them pairs of point-tracks.

The first axiom of congruence is that the opposite sides of any parallelogram are congruent. This axiom enables us to compare the lengths of any two segments either respectively on parallel rects or on the same rect. Also it enables us to compare the lengths of any two segments either respectively on parallel point-tracks or on the same point-track. It follows from this axiom that two objects at rest in any two points of a time-system β are moving with equal velocities in any other time-system α along parallel lines. Thus we can speak of the velocity in α due to the time-system β without specifying any particular point in β. The axiom also enables us to measure time in any time-system; but does not enable us to compare times in different time-systems.

The second axiom of congruence concerns parallelograms on congruent bases and between the same parallels, which have also their other pairs of sides parallel. The axiom asserts that the rect joining the two event-particles of intersection of the diagonals is parallel to the rect on which the bases lie. By the aid of this axiom it easily follows that the diagonals of a parallelogram bisect each other.

Congruence is extended in any space beyond parallel rects to all rects by two axioms depending on perpendicularity. The first of these axioms, which is the third axiom of congruence, is that if ABC is a triangle of rects in any moment and D is the middle event-particle of the base BC, then the level through D perpendicular to BC contains A when and only when AB is congruent to AC. This axiom evidently expresses the symmetry of perpendicularity, and is the essence of the famous pons asinorum expressed as an axiom.

The second axiom depending on perpendicularity,

and the fourth axiom of congruence, is that if r and A be a rect and an event-particle in the same moment and AB and AC be a pair of rectangular rects intersecting r in B and C, and AD and AE be another pair of rectangular rects intersecting r in D and E, then either D or E lies in the segment BC and the other one of the two does not lie in this segment. Also as a particular case of this axiom, if AB be perpendicular to r and in consequence AC be parallel to r, then D and E lie on opposite sides of B respectively. By the aid of these two axioms the theory of congruence can be extended so as to compare lengths of segments on any two rects. Accordingly Euclidean metrical geometry in space is completely established and lengths in the spaces of different time-systems are comparable as the result of definite properties of nature which indicate just that particular method of comparison.

The comparison of time-measurements in diverse time-systems requires two other axioms. The first of these axioms, forming the fifth axiom of congruence, will be called the axiom of 'kinetic symmetry.' It expresses the symmetry of the quantitative relations between two time-systems when the times and lengths in the two systems are measured in congruent units.

The axiom can be explained as follows: Let α and β be the names of two time-systems. The directions of motion in the space of α due to rest in a point of β is called the 'β-direction in α' and the direction of motion in the space of β due to rest in a point of α is called the 'α-direction in β.' Consider a motion in the space of α consisting of a certain velocity in the β-direction of α and a certain velocity at right-angles to it. This motion represents rest in the space of another time-system—

call it π. Rest in π will also be represented in the space of β by a certain velocity in the α-direction in β and a certain velocity at right-angles to this α-direction. Thus a certain motion in the space of α is correlated to a certain motion in the space of β, as both representing the same fact which can also be represented by rest in π. Now another time-system, which I will name σ, can be found which is such that rest in its space is represented by the same magnitudes of velocities along and perpendicular to the α-direction in β as those velocities in α, along and perpendicular to the β-direction, which represent rest in π. The required axiom of kinetic symmetry is that rest in σ will be represented in α by the same velocities along and perpendicular to the β-direction in α as those velocities in β along and perpendicular to the α-direction which represent rest in π.

A particular case of this axiom is that relative velocities are equal and opposite. Namely rest in α is represented in β by a velocity along the α-direction which is equal to the velocity along the β-direction in α which represents rest in β.

Finally the sixth axiom of congruence is that the relation of congruence is transitive. So far as this axiom applies to space, it is superfluous. For the property follows from our previous axioms. It is however necessary for time as a supplement to the axiom of kinetic symmetry. The meaning of the axiom is that if the time-unit of system α is congruent to the time-unit of system β, and the time-unit of system β is congruent to the time-unit of system γ, then the time-units of α and γ are also congruent.

By means of these axioms formulae for the transformation of measurements made in one time-system to measurements of the same facts of nature made in another time-system can be deduced. These formulae will be found to involve one arbitrary constant which I will call k.

It is of the dimensions of the square of a velocity. Accordingly four cases arise. In the first case k is zero. This case produces nonsensical results in opposition to the elementary deliverances of experience. We put this case aside.

In the second case k is infinite. This case yields the ordinary formulae for transformation in relative motion, namely those formulae which are to be found in every elementary book on dynamics.

In the third case, k is negative. Let us call it $-c^2$, where c will be of the dimensions of a velocity. This case yields the formulae of transformation which Larmor discovered for the transformation of Maxwell's equations of the electromagnetic field. These formulae were extended by H. A. Lorentz, and used by Einstein and Minkowski as the basis of their novel theory of relativity. I am not now speaking of Einstein's more recent theory of general relativity by which he deduces his modification of the law of gravitation. If this be the case which applies to nature, then c must be a close approximation to the velocity of light *in vacuo*. Perhaps it is this actual velocity. In this connexion '*in vacuo*' must not mean an absence of events, namely the absence of the all-pervading ether of events. It must mean the absence of certain types of objects.

In the fourth case, k is positive. Let us call it h^2, where h will be of the dimensions of a velocity. This gives a perfectly possible type of transformation formulae, but not one which explains any facts of experience. It has also another disadvantage. With the assumption of this fourth case the distinction between space and time becomes unduly blurred. The whole object of these lectures has been to enforce the doctrine that space and time spring from a common root, and that the ultimate fact of experience is a space-time fact. But after all mankind does distinguish very sharply between space and time, and it is owing to this sharpness of distinction that the doctrine of these lectures is somewhat of a paradox. Now in the third assumption this sharpness of distinction is adequately preserved. There is a fundamental distinction between the metrical properties of point-tracks and rects. But in the fourth assumption this fundamental distinction vanishes.

Neither the third nor the fourth assumption can agree with experience unless we assume that the velocity c of the third assumption, and the velocity h of the fourth assumption, are extremely large compared to the velocities of ordinary experience. If this be the case the formulae of both assumptions will obviously reduce to a close approximation to the formulae of the second assumption which are the ordinary formulae of dynamical textbooks. For the sake of a name, I will call these textbook formulae the 'orthodox' formulae.

There can be no question as to the general approximate correctness of the orthodox formulae. It would be merely silly to raise doubts on this point. But the determination of the status of these formulae is by no means settled by this admission. The independence of time and space is an unquestioned presupposition of the orthodox thought which has produced the orthodox formulae. With this presupposition and given the absolute points of one absolute space, the orthodox formulae are immediate deductions. Accordingly, these formulae are presented to our imaginations as facts which cannot be otherwise, time and space being what they are. The orthodox formulae have therefore attained to the status of necessities which cannot be questioned in science. Any attempt to replace these formulae by others was to abandon the *rôle* of physical explanation and to have recourse to mere mathematical formulae.

But even in physical science difficulties have accumulated round the orthodox formulae. In the first place Maxwell's equations of the electromagnetic field are not invariant for the transformations of the orthodox formulae; whereas they are invariant for the transformations of the formulae arising from the third of the four cases mentioned above, provided that the velocity c is identified with a famous electromagnetic constant quantity.

Again the null results of the delicate experiments to detect the earth's variations of motion through the ether in its orbital path are explained immediately by the formulae of the third case. But if we assume the orthodox formulae we have to make a special and arbitrary assumption as to the contraction of matter during motion. I mean the Fitzgerald-Lorentz assumption.

Lastly Fresnel's coefficient of drag which represents the variation of the velocity of light in a moving medium is explained by the formulae of the third case, and requires another arbitrary assumption if we use the orthodox formulae.

It appears therefore that on the mere basis of physical explanation there are advantages in the formulae of the third case as compared with the orthodox formulae. But the way is blocked by the ingrained belief that these latter formulae possess a character of necessity. It is therefore an urgent requisite for physical science and for philosophy to examine critically the grounds for this supposed necessity. The only satisfactory method of scrutiny is to recur to the first principles of our knowledge of nature. This is exactly what I am endeavouring to do in these lectures. I ask what it is that we are aware of in our sense-perception of nature. I then proceed to examine those factors in nature which lead us to conceive nature as occupying space and persisting through time. This procedure has led us to an investigation of the characters of space and time. It results from these investigations that the formulae of the third case and the orthodox formulae are on a level as possible formulae resulting from the basic character of our knowledge of nature. The orthodox formulae have thus lost any advantage as to necessity which they enjoyed over the serial group. The way is thus open to adopt whichever of the two groups best accords with observation.

I take this opportunity of pausing for a moment from the course of my argument, and of reflecting on the general character which my doctrine ascribes to some familiar concepts of science. I have no doubt that some of you have felt that in certain aspects this character is very paradoxical.

This vein of paradox is partly due to the fact that educated language has been made to conform to the prevalent orthodox theory. We are thus, in expounding an alternative doctrine, driven to the use of either strange terms or of familiar words with unusual meanings. This victory of the orthodox theory over language is very natural. Events are named after the prominent objects situated in them, and thus both in language and in thought the event sinks behind the object, and becomes the mere play of its relations. The theory of space is then converted into a theory of the relations of objects instead of a theory of the relations of events. But objects have not the passage of events. Accordingly space as a relation between objects is devoid of any connexion with time. It is space at an instant without any determinate relations between the spaces at successive instants. It cannot be one timeless space because the relations between objects change.

A few minutes ago in speaking of the deduction of the orthodox formulae for relative motion I said that they followed as an immediate deduction from the assumption of absolute points in absolute space. This reference to absolute space was not an oversight. I know that the doctrine of the relativity of space at present holds the field both in science and philosophy. But I do not think that its inevitable consequences are understood. When we really face them the paradox of the presentation of the character of space which I have elaborated is greatly mitigated. If there is no absolute position, a point must cease to be a simple entity. What is a point to one man in a balloon with his eyes fixed on an instrument is a track of points to an observer on the earth who is watching the balloon through a telescope, and is another track of points to an observer in the sun who is watching the balloon through some instrument suited to such a being. Accordingly if I am reproached with the paradox of my theory of points as classes of event-particles, and of my theory of event-particles as groups of abstractive sets, I ask my critic to explain exactly what he means by a point. While you explain your meaning about anything, however simple, it is always apt to look subtle and fine spun. I have at least explained exactly what I do mean by a point, what relations it involves and what entities are the relata. If you admit the relativity of space, you also must admit that points are complex entities, logical constructs involving other entities and their relations. Produce your theory, not in a few vague phrases of indefinite meaning, but explain it step by step in definite terms referring to assigned relations and assigned relata. Also show that your theory of points issues in a theory of space. Furthermore note that the example of the man in the balloon, the observer on earth, and the observer in the sun, shows that every assumption of relative rest requires a timeless space with radically different points from those which issue from every other such assumption. The theory of the relativity of space is inconsistent with any doctrine of one unique set of points of one timeless space.

The fact is that there is no paradox in my doctrine of the nature of space which is not in essence inherent in the theory of the relativity of space. But this doctrine has never really been accepted in science, whatever people say. What appears in our dynamical treatises is Newton's doctrine of relative motion based on the doctrine of differential motion in absolute space. When you once admit that the points are radically different entities for differing assumptions of rest, then the orthodox formulae lose all their obviousness. They were only obvious because you were really thinking of something else. When discussing this topic you can only avoid paradox by taking refuge from the flood of criticism in the comfortable ark of no meaning.

The new theory provides a definition of the congruence of periods of time. The prevalent view provides no such definition. Its position is that if we take such time-measurements so that certain familiar velocities which seem to us to be uniform are uniform, then the laws of motion are true. Now in the first place no change could appear either as uniform or non-uniform without involving a definite determination of the congruence for time-periods. So in appealing to familiar phenomena it allows that there is some factor in nature which we can intellectually construct as a congruence theory. It does not however say anything about it except that the laws of motion are then true. Suppose that with some expositors we cut out the reference to familiar velocities such as the rate of rotation of the earth. We are then driven to admit that there is no meaning in temporal congruence except that certain assumptions make the laws of motion true. Such a statement is historically false. King Alfred the Great was ignorant of the laws of motion, but knew very well what he meant by the measurement of time, and achieved his purpose by means of burning candles. Also no one in past ages justified the use of sand in hour-glasses by saying that some centuries later interesting laws of motion would be discovered which would give a meaning to the statement that the sand was emptied from the bulbs in equal times. Uniformity in change is directly perceived, and it follows that mankind perceives in nature factors from which a theory of temporal congruence can be formed. The prevalent theory entirely fails to produce such factors.

The mention of the laws of motion raises another point where the prevalent theory has nothing to say and the new theory gives a complete explanation. It is well known that the laws of motion are not valid for any axes of reference which you may choose to take fixed in any rigid body. You must choose a body which is not rotating and has no acceleration. For example they do not really apply to axes fixed in the earth because of the diurnal rotation of that body. The law which fails when you assume the wrong axes as at rest is the third law, that action and reaction are equal and opposite. With the wrong axes uncompensated centrifugal forces and uncompensated composite centrifugal forces appear, due to rotation. The influence of these forces can be demonstrated by many facts on the earth's surface, Foucault's pendulum, the shape of the earth, the fixed directions of the rotations of cyclones and anticyclones. It is difficult to take seriously the suggestion that these domestic phenomena on the earth are due to the influence of the fixed stars. I cannot persuade myself to believe that a little star in its twinkling turned round Foucault's pendulum in the Paris Exhibition of 1861. Of course anything is believable when a definite physical connexion has been demonstrated, for example the influence of sunspots. Here all demonstration is lacking in the form of any coherent theory. According to the theory of these lectures the axes to which motion is to be referred are axes at rest in the space of some time-system. For example, consider the space of a time-system α. There are sets of axes at rest in the space of α. These are suitable dynamical axes. Also a set of axes in this space which is moving with uniform velocity without rotation is another suitable set. All the moving points fixed in these moving axes are really tracing out parallel lines with one uniform velocity. In other words they are the reflections in the space of α of a set of fixed axes in the space of some other time-system β. Accordingly the group of dynamical axes required for Newton's Laws of Motion is the outcome of the necessity of referring motion to a body at rest in the space of some one time-system in order to obtain a coherent account of physical properties. If we do not do so the meaning of the motion of one portion of our physical configuration is different from the meaning of the motion of another portion of the same configuration. Thus the meaning of motion being what it is, in order to describe the motion of any system of objects without changing the meaning of your terms as you proceed with your description, you are bound to take one of these sets of axes as axes of reference; though you may choose their reflections into the space of any time-system which you wish to adopt. A definite physical reason is thereby assigned for the peculiar property of the dynamical group of axes.

On the orthodox theory the position of the equations of motion is most ambiguous. The space to which they refer is completely undetermined and so is the measurement of the lapse of time. Science is simply setting out on a fishing expedition to see whether it cannot find some procedure which it can call the measurement of space and some procedure which it can call the measurement of time, and something which it can call a system of forces, and something which it can call masses, so that these formulae may be satisfied. The only reason—on this theory—why anyone should want to satisfy these formulae is a sentimental regard for Galileo,

Newton, Euler and Lagrange. The theory, so far from founding science on a sound observational basis, forces everything to conform to a mere mathematical preference for certain simple formulae.

I do not for a moment believe that this is a true account of the real status of the Laws of Motion. These equations want some slight adjustment for the new formulae of relativity. But with these adjustments, imperceptible in ordinary use, the laws deal with fundamental physical quantities which we know very well and wish to correlate.

The measurement of time was known to all civilised nations long before the laws were thought of. It is this time as thus measured that the laws are concerned with. Also they deal with the space of our daily life. When we approach to an accuracy of measurement beyond that of observation, adjustment is allowable. But within the limits of observation we know what we mean when we speak of measurements of space and measurements of time and uniformity of change. It is for science to give an intellectual account of what is so evident in sense-awareness. It is to me thoroughly incredible that the ultimate fact beyond which there is no deeper explanation is that mankind has really been swayed by an unconscious desire to satisfy the mathematical formulae which we call the Laws of Motion, formulae completely unknown till the seventeenth century of our epoch.

The correlation of the facts of sense-experience effected by the alternative account of nature extends beyond the physical properties of motion and the properties of congruence. It gives an account of the meaning of the geometrical entities such as points, straight lines, and volumes, and connects the kindred ideas of extension in time and extension in space. The theory satisfies the true purpose of an intellectual explanation in the sphere of natural philosophy. This purpose is to exhibit the interconnexions of nature, and to show that one set of ingredients in nature requires for the exhibition of its character the presence of the other sets of ingredients.

The false idea which we have to get rid of is that of nature as a mere aggregate of independent entities, each capable of isolation. According to this conception these entities, whose characters are capable of isolated definition, come together and by their accidental relations form the system of nature. This system is thus thoroughly accidental; and, even if it be subject to a mechanical fate, it is only accidentally so subject.

With this theory space might be without time, and time might be without space. The theory admittedly breaks down when we come to the relations of matter and space. The relational theory of space is an admission that we cannot know space without matter or matter without space. But the seclusion of both from time is still jealously guarded. The relations between portions of matter in space are accidental facts owing to the absence of any coherent account of how space springs from matter or how matter springs from space. Also what we really observe in nature, its colours and its sounds and its touches are secondary qualities; in other words, they are not in nature at all but are accidental products of the relations between nature and mind.

The explanation of nature which I urge as an alternative ideal to this accidental view of nature, is that nothing in nature could be what it is except as an ingredient in nature as it is. The whole which is present for discrimination is posited in sense-awareness as necessary for the discriminated parts. An isolated event is not an event, because every event is a factor in a larger whole and is significant of that whole. There can be no time apart from space; and no space apart from time; and no space and no time apart from the passage of the events of nature. The isolation of an entity in thought, when we think of it as a bare 'it,' has no counterpart in any corresponding isolation in nature. Such isolation is merely part of the procedure of intellectual knowledge.

The laws of nature are the outcome of the characters of the entities which we find in nature. The entities being what they are, the laws must be what they are; and conversely the entities follow from the laws. We are a long way from the attainment of such an ideal; but it remains as the abiding goal of theoretical science.

CHAPTER VII

OBJECTS

THE ensuing lecture is concerned with the theory of objects. Objects are elements in nature which do not pass. The awareness of an object as some factor not sharing in the passage of nature is what I call 'recognition.' It is impossible to recognise an event, because an event is essentially distinct from every other event. Recognition is an awareness of sameness. But to call recognition an awareness of sameness implies an intellectual act of comparison accompanied with judgment. I use recognition for the non-intellectual relation of sense-awareness which connects the mind with a factor of nature without passage. On the intellectual side of the mind's experience there are comparisons of things recognised and consequent judgments of sameness or diversity. Probably 'sense-recognition' would be a better term for what I mean by 'recognition.' I have chosen the simpler term because I think that I shall be able to avoid the use of 'recognition' in any other meaning than that of 'sense-recognition.' I am quite willing to believe that recognition, in my sense of the term, is merely an ideal limit, and that there is in fact no recognition without intellectual accompaniments of comparison and judgment. But recognition is that relation of the mind to nature which provides the material for the intellectual activity.

An object is an ingredient in the character of some event. In fact the character of an event is nothing but the objects which are ingredient in it and the ways in which those objects make their ingression into the event. Thus the theory of objects is the theory of the comparison of events. Events are only comparable because they body forth permanences. We are comparing objects in events whenever we can say, 'There it is again.' Objects are the elements in nature which can 'be again.'

Sometimes permanences can be proved to exist which evade recognition in the sense in which I am using that term. The permanences which evade recognition appear to us as abstract properties either of events or of objects. All the same they are there for recognition although undiscriminated in our sense-awareness. The demarcation of events, the splitting of nature up into parts is effected by the objects which we recognise as their ingredients. The discrimination of nature is the recognition of objects amid passing events. It is a compound of the awareness of the passage of nature, of the consequent partition of nature, and of the definition of certain parts of nature by the modes of the ingression of objects into them.

You may have noticed that I am using the term 'ingression' to denote the general relation of objects to events. The ingression of an object into an event is the way the character of the event shapes itself in virtue of the being of the object. Namely the event is what it is, because the object is what it is; and when I am thinking of this modification of the event by the object, I call the relation between the two 'the ingression of the object into the event.' It is equally true to say that objects are what they are because events are what they are. Nature is such that there can be no events and no objects without the ingression of objects into events.

Although there are events such that the ingredient objects evade our recognition. These are the events in empty space. Such events are only analysed for us by the intellectual probing of science.

Ingression is a relation which has various modes. There are obviously very various kinds of objects; and no one kind of object can have the same sort of relations to events as objects of another kind can have. We shall have to analyse out some of the different modes of ingression which different kinds of objects have into events.

But even if we stick to one and the same kind of objects, an object of that kind has different modes of ingression into different events. Science and philosophy have been apt to entangle themselves in a simple-minded theory that an object is at one place at any definite time, and is in no sense anywhere else. This is in fact the attitude of common sense thought, though it is not the attitude of language which is naïvely expressing the facts of experience. Every other sentence in a work of literature which is endeavouring truly to interpret the facts of experience expresses differences in surrounding events due to the presence of some object. An object is ingredient throughout its neighbourhood, and its neighbourhood is indefinite. Also the modification of events by ingression is susceptible of quantitative differences. Finally therefore we are driven to admit that each object is in some sense ingredient throughout nature; though its ingression may be quantitatively irrelevant in the expression of our individual experiences.

This admission is not new either in philosophy or science. It is obviously a necessary axiom for those philosophers who insist that reality is a system. In these lectures we are keeping off the profound and vexed question as to what we mean by 'reality.' I am maintaining the humbler thesis that nature is a system. But I suppose that in this case the less follows from the greater, and that I may claim the support of these philosophers. The same doctrine is essentially interwoven in all modern physical speculation. As long ago as 1847 Faraday in a paper in the *Philosophical Magazine* remarked that his theory of tubes of force implies that in a sense an electric charge is everywhere. The modification of the electromagnetic field at every point of space at each instant owing to the past history of each electron is another way of stating the same fact. We can however illustrate the doctrine by the more familiar facts of life without recourse to the abstruse speculations of theoretical physics.

The waves as they roll on to the Cornish coast tell of a gale in mid-Atlantic; and our dinner witnesses to the ingression of the cook into the dining room. It is evident that the ingression of objects into events includes the theory of causation. I prefer to neglect this aspect of ingression, because causation raises the memory of discussions based upon theories of nature which are alien to my own. Also I think that some new light may be thrown on the subject by viewing it in this fresh aspect.

The examples which I have given of the ingression of objects into events remind us that ingression takes a peculiar form in the case of some events; in a sense, it is a more concentrated form. For example, the electron has a certain position in space and a certain shape. Perhaps it is an extremely small sphere in a certain test-tube. The storm is a gale situated in mid-Atlantic with a certain latitude and longitude, and the cook is in the kitchen. I will call this special form of ingression the 'relation of situation'; also, by a double use of the word 'situation,' I will call the event in which an object is situated 'the situation of the object.' Thus a situation is an event which is a relatum in the relation of situation. Now our first impression is that at last we have come to the simple plain fact of where the object really is; and that the vaguer relation which I call ingression should not be muddled up with the relation of situation, as if including it as a particular case. It seems so obvious that any object is in such and such a position, and that it is influencing other events in a totally different sense. Namely, in a sense an object is the character of the event which is its situation, but it only influences the character of other events. Accordingly the relations of situation and influencing are not generally the same sort of relation, and should not be subsumed under the same term 'ingression.' I believe that this notion is a mistake, and that it is impossible to draw a clear distinction between the two relations.

For example, Where was your toothache? You went to a dentist and pointed out the tooth to him. He pronounced it perfectly sound, and cured you by stopping another tooth. Which tooth was the situation of the toothache? Again, a man has an arm amputated, and experiences sensations in the hand which he has lost. The situation of the imaginary hand is in fact merely thin air. You look into a mirror and see a fire. The flames that you see are situated behind the mirror. Again at night you watch the sky; if some of the stars had vanished from existence hours ago, you would not be any the wiser. Even the situations of the planets differ from those which science would assign to them.

Anyhow you are tempted to exclaim, the cook is in the kitchen. If you mean her mind, I will not agree with you on the point; for I am only talking of nature. Let us think only of her bodily presence. What do you mean by this notion? We confine ourselves to typical manifestations of it. You can see her, touch her, and hear her. But the examples which I have given you show that the notions of the situations of what you see, what you touch, and what you hear are not so sharply separated out as to defy further questioning. You cannot cling to the idea that we have two sets of experiences of nature, one of primary qualities which belong to the objects perceived, and one of secondary qualities which are the products of our mental excitements. All we know of nature is in the same boat, to sink or swim together. The constructions of science are merely expositions of the characters of things perceived. Accordingly to affirm that the cook is a certain dance of molecules and electrons is merely to affirm that the things about her which are perceivable have certain characters. The situations of the perceived manifestations of her bodily presence have only a very general relation to the situations of the molecules, to be determined by discussion of the circumstances of perception.

In discussing the relations of situation in particular and of ingression in general, the first requisite is to note that objects are of radically different types. For each type 'situation' and 'ingression' have their own special meanings which are different from their meanings for other types, though connexions can be pointed out.

It is necessary therefore in discussing them to determine what type of objects are under consideration. There are, I think, an indefinite number of types of objects. Happily we need not think of them all. The idea of situation has its peculiar importance in reference to three types of objects which I call sense-objects, perceptual objects and scientific objects. The suitability of these names for the three types is of minor importance, so long as I can succeed in explaining what I mean by them.

These three types form an ascending hierarchy, of which each member presupposes the type below. The base of the hierarchy is formed by the sense-objects. These objects do not presuppose any other type of objects. A sense-object is a factor of nature posited by sense-awareness which (i), in that it is an object, does not share in the passage of nature and (ii) is not a relation between other factors of nature. It will of course be a relatum in relations which also implicate other factors of nature. But it is always a relatum and never the relation itself. Examples of sense-objects are a particular sort of colour, say Cambridge blue, or a particular sort of sound, or a particular sort of smell, or a particular sort of feeling. I am not talking of a particular patch of blue as seen during a particular second of time at a definite date. Such a patch is an event where Cambridge blue is situated. Similarly I am not talking of any particular concert-room as filled with the note. I mean the note itself and not the patch of volume filled by the sound for a tenth of a second. It is natural for us to think of the note in itself, but in the case of colour we are apt to think of it merely as a property of the patch. No one thinks of the note as a property of the concert-room. We see the blue and we hear the note. Both the blue and the note are immediately posited by the discrimination of sense-awareness which relates the mind to nature. The blue is posited as in nature related to other factors in nature. In particular it is posited as in the relation of being situated in the event which is its situation.

The difficulties which cluster around the relation of situation arise from the obstinate refusal of philosophers to take seriously the ultimate fact of multiple relations. By a multiple relation I mean a relation which in any concrete instance of its occurrence necessarily involves more than two relata. For example, when John likes Thomas there are only two relata, John and Thomas. But when John gives that book to Thomas there are three relata, John, that book, and Thomas.

Some schools of philosophy, under the influence of the Aristotelian logic and the Aristotelian philosophy, endeavour to get on without admitting any relations at all except that of substance and attribute. Namely all apparent relations are to be resolvable into the concurrent existence of substances with contrasted attributes. It is fairly obvious that the Leibnizian monadology is the necessary outcome of any such philosophy. If you dislike pluralism, there will be only one monad.

Other schools of philosophy admit relations but obstinately refuse to contemplate relations with more than two relata. I do not think that this limitation is based on any set purpose or theory. It merely arises from the fact that more complicated relations are a bother to people without adequate mathematical training, when they are admitted into the reasoning.

I must repeat that we have nothing to do in these

lectures with the ultimate character of reality. It is quite possible that in the true philosophy of reality there are only individual substances with attributes, or that there are only relations with pairs of relata. I do not believe that such is the case; but I am not concerned to argue about it now. Our theme is Nature. So long as we confine ourselves to the factors posited in the sense-awareness of nature, it seems to me that there certainly are instances of multiple relations between these factors, and that the relation of situation for sense-objects is one example of such multiple relations.

Consider a blue coat, a flannel coat of Cambridge blue belonging to some athlete. The coat itself is a perceptual object and its situation is not what I am talking about. We are talking of someone's definite sense-awareness of Cambridge blue as situated in some event of nature. He may be looking at the coat directly. He then sees Cambridge blue as situated practically in the same event as the coat at that instant. It is true that the blue which he sees is due to light which left the coat some inconceivably small fraction of a second before. This difference would be important if he were looking at a star whose colour was Cambridge blue. The star might have ceased to exist days ago, or even years ago. The situation of the blue will not then be very intimately connected with the situation (in another sense of 'situation') of any perceptual object. This disconnexion of the situation of the blue and the situation of some associated perceptual object does not require a star for its exemplification. Any looking glass will suffice. Look at the coat through a looking glass. Then blue is seen as situated behind the mirror. The event which is its situation depends upon the position of the observer.

The sense-awareness of the blue as situated in a certain event which I call the situation, is thus exhibited as the sense-awareness of a relation between the blue, the percipient event of the observer, the situation, and intervening events. All nature is in fact required, though only certain intervening events require their characters to be of certain definite sorts. The ingression of blue into the events of nature is thus exhibited as systematically correlated. The awareness of the observer depends on the position of the percipient event in this systematic correlation. I will use the term 'ingression into nature' for this systematic correlation of the blue with nature. Thus the ingression of blue into any definite event is a part statement of the fact of the ingression of blue into nature.

In respect to the ingression of blue into nature events may be roughly put into four classes which overlap and are not very clearly separated. These classes are (i) the percipient events, (ii) the situations, (iii) the active conditioning events, (iv) the passive conditioning events. To understand this classification of events in the general fact of the ingression of blue into nature, let us confine attention to one situation for one percipient event and to the consequent *rôles* of the conditioning events for the ingression as thus limited. The percipient event is the relevant bodily state of the observer. The situation is where he sees the blue, say, behind the mirror. The active conditioning events are the events whose characters are particularly relevant for the event (which is the situation) to be the situation for that percipient event, namely the coat, the mirror, and the state of the room as to light and atmosphere. The passive conditioning events are the events of the rest of nature.

In general the situation is an active conditioning event; namely the coat itself, when there is no mirror or other such contrivance to produce abnormal effects. But the example of the mirror shows us that the situation may be one of the passive conditioning events. We are then apt to say that our senses have been cheated, because we demand as a right that the situation should be an active condition in the ingression.

This demand is not so baseless as it may seem when presented as I have put it. All we know of the characters of the events of nature is based on the analysis of the relations of situations to percipient events. If situations were not in general active conditions, this analysis would tell us nothing. Nature would be an unfathomable enigma to us and there could be no science. Accordingly the incipient discontent when a situation is found to be a passive condition is in a sense justifiable; because if that sort of thing went on too often, the *rôle* of the intellect would be ended.

Furthermore the mirror is itself the situation of other sense-objects either for the same observer with the same percipient event, or for other observers with other percipient events. Thus the fact that an event is a situation in the ingression of one set of sense-objects into nature is presumptive evidence that that event is an active condition in the ingression of other sense-objects into nature which may have other situations.

This is a fundamental principle of science which it has derived from common sense.

I now turn to perceptual objects. When we look at the coat, we do not in general say, There is a patch of Cambridge blue; what naturally occurs to us is, There is a coat. Also the judgment that what we have seen is a garment of man's attire is a detail. What we perceive is an object other than a mere sense-object. It is not a mere patch of colour, but something more; and it is that something more which we judge to be a coat. I will use the word 'coat' as the name for that crude object which is more than a patch of colour, and without any allusion to the judgments as to its usefulness as an article of attire either in the past or the future. The coat which is perceived—in this sense of the word 'coat'—is what I call a perceptual object. We have to investigate the general character of these perceptual objects.

It is a law of nature that in general the situation of a sense-object is not only the situation of that sense-object for one definite percipient event, but is the situation of a variety of sense-objects for a variety of percipient events. For example, for any one percipient event, the situation of a sense-object of sight is apt also to be the situations of sense-objects of sight, of touch, of smell, and of sound. Furthermore this concurrence in the situations of sense-objects has led to the body—*i.e.* the percipient event—so adapting itself that the perception of one sense-object in a certain situation leads to a subconscious sense-awareness of other sense-objects in the same situation. This interplay is especially the case between touch and sight. There is a certain correlation between the ingressions of sense-objects of touch and sense-objects of sight into nature, and in a slighter degree between the ingressions of other pairs of sense-objects. I call this sort of correlation the 'conveyance' of one sense-object by another. When you see the blue flannel coat you subconsciously feel yourself wearing it or otherwise touching it. If you are a smoker, you may also subconsciously be aware of the faint aroma of tobacco. The peculiar fact, posited by this sense-awareness of the concurrence of subconscious sense-objects along with one or more dominating sense-objects in the same situation, is the sense-awareness of the perceptual object. The perceptual object is not primarily the issue of a judgment. It is a factor of nature directly posited in sense-awareness. The element of judgment comes in when we proceed to classify the particular perceptual object. For example, we say, That is flannel, and we think of the properties of flannel and the uses of athletes' coats. But that all takes place after we have got hold of the perceptual object. Anticipatory judgments affect the perceptual object perceived by focussing and diverting attention.

The perceptual object is the outcome of the habit of experience. Anything which conflicts with this habit hinders the sense-awareness of such an object. A sense-object is not the product of the association of intellectual ideas; it is the product of the association of sense-objects in the same situation. This outcome is not intellectual; it is an object of peculiar type with its own particular ingression into nature.

There are two kinds of perceptual objects, namely, 'delusive perceptual objects' and 'physical objects.' The situation of a delusive perceptual object is a passive condition in the ingression of that object into nature. Also the event which is the situation will have the relation of situation to the object only for one particular percipient event. For example, an observer sees the image of the blue coat in a mirror. It is a blue coat that he sees and not a mere patch of colour. This shows that the active conditions for the conveyance of a group of subconscious sense-objects by a dominating sense-object are to be found in the percipient event. Namely we are to look for them in the investigations of medical psychologists. The ingression into nature of the delusive sense-object is conditioned by the adaptation of bodily events to the more normal occurrence, which is the ingression of the physical object.

A perceptual object is a physical object when (i) its situation is an active conditioning event for the ingression of any of its component sense-objects, and (ii) the same event can be the situation of the perceptual object for an indefinite number of possible percipient events. Physical objects are the ordinary objects which we perceive when our senses are not cheated, such as chairs, tables and trees. In a way physical objects have more insistent perceptive power than sense-objects. Attention to the fact of their occurrence in nature is the first condition for the survival of complex living organisms. The result of this high perceptive power of physical objects is the scholastic philosophy of nature which looks on the sense-objects as mere attributes of the physical objects. This scholastic point of view is directly contradicted by the wealth of sense-objects which enter into our experience as situated in events without any connexion with physical objects. For example, stray smells, sounds, colours and more subtle nameless sense-objects. There is no perception of physical objects without perception of sense-objects. But the converse does not hold: namely, there is abundant perception of sense-objects unaccompanied by any perception of physical objects. This lack of reciprocity in the relations between sense-objects and physical objects is fatal to the scholastic natural philosophy.

There is a great difference in the *rôles* of the situations of sense-objects and physical objects. The situations of a physical object are conditioned by uniqueness and continuity. The uniqueness is an ideal limit to which we approximate as we proceed in thought along an abstractive set of durations, considering smaller and smaller durations in the approach to the ideal limit of the moment of time. In other words, when the duration is small enough, the situation of the physical object within that duration is practically unique.

The identification of the same physical object as being situated in distinct events in distinct durations is effected by the condition of continuity. This condition of continuity is the condition that a continuity of passage of events, each event being a situation of the object in its corresponding duration, can be found from the earlier to the later of the two given events. So far as the two events are practically adjacent in one specious present, this continuity of passage may be directly perceived. Otherwise it is a matter of judgment and inference.

The situations of a sense-object are not conditioned by any such conditions either of uniqueness or of continuity. In any durations however small a sense-object may have any number of situations separated from each other. Thus two situations of a sense-object, either in the same duration or in different durations, are not necessarily connected by any continuous passage of events which are also situations of that sense-object.

The characters of the conditioning events involved in the ingression of a sense-object into nature can be largely expressed in terms of the physical objects which are situated in those events. In one respect this is also a tautology. For the physical object is nothing else than the habitual concurrence of a certain set of sense-objects in one situation. Accordingly when we know all about the physical object, we thereby know its component sense-objects. But a physical object is a condition for the occurrence of sense-objects other than those which are its components. For example, the atmosphere causes the events which are its situations to be active conditioning events in the transmission of sound. A mirror which is itself a physical object is an active condition for the situation of a patch of colour behind it, due to the reflection of light in it.

Thus the origin of scientific knowledge is the endeavour to express in terms of physical objects the various *rôles* of events as active conditions in the ingression of sense-objects into nature. It is in the progress of this investigation that scientific objects emerge. They embody those aspects of the character of the situations of the physical objects which are most permanent and are expressible without reference to a multiple relation including a percipient event. Their relations to each other are also characterised by a certain simplicity and uniformity. Finally the characters of the observed physical objects and sense-objects can be expressed in terms of these scientific objects. In fact the whole point of the search for scientific objects is the endeavour to obtain this simple expression of the characters of events. These scientific objects are not themselves merely formulae for calculation; because formulae must refer to things in nature, and the scientific objects are the things in nature to which the formulae refer.

A scientific object such as a definite electron is a systematic correlation of the characters of all events throughout all nature. It is an aspect of the systematic character of nature. The electron is not merely where its charge is. The charge is the quantitative character of certain events due to the ingression of the electron into nature. The electron is its whole field of force. Namely the electron is the systematic way in which all events are modified as the expression of its ingression. The situation of an electron in any small duration may be defined as that event which has the quantitative character which is the charge of the electron. We may if we please term the mere charge the electron. But then another name is required for the scientific object which is the full entity which concerns science, and which I have called the electron.

According to this conception of scientific objects, the rival theories of action at a distance and action by transmission through a medium are both incomplete expressions of the true process of nature. The stream of events which form the continuous series of situations of the electron is entirely self-determined, both as regards having the intrinsic character of being the series of situations of that electron and as regards the time-systems with which its various members are cogredient, and the flux of their positions in their corresponding durations. This is the foundation of the denial of action at a distance; namely the progress of the stream of the situations of a scientific object can be determined by an analysis of the stream itself.

On the other hand the ingression of every electron into nature modifies to some extent the character of every event. Thus the character of the stream of events which we are considering bears marks of the existence of every other electron throughout the universe. If we like to think of the electrons as being merely what I call their charges, then the charges act at a distance. But this action consists in the modification of the situation of the other electron under consideration. This conception of a charge acting at a distance is a wholly artificial one. The conception which most fully expresses the character of nature is that of each event as modified by the ingression of each electron into nature. The ether is the expression of this systematic modification of events throughout space and throughout time. The best expression of the character of this modification is for physicists to find out. My theory has nothing to do with that and is ready to accept any outcome of physical research.

The connexion of objects with space requires elucidation. Objects are situated in events. The relation of situation is a different relation for each type of object, and in the case of sense-objects it cannot be expressed as a two-termed relation. It would perhaps be better to use a different word for these different types of the relation of situation. It has not however been necessary to do so for our purposes in these lectures. It must be understood however that, when situation is spoken of, some one definite type is under discussion, and it may happen that the argument may not apply to situation of another type. In all cases however I use situation to express a relation between objects and events and not between objects and abstractive elements. There is a derivative relation between objects and spatial elements which I call the relation of location; and when this relation holds, I say that the object is located in the abstractive element. In this sense, an object may be located in a moment of time, in a volume of space, an area, a line, or a point. There will be a peculiar type of location corresponding to each type of situation; and location is in each case derivative from the corresponding relation of situation in a way which I will proceed to explain.

Also location in the timeless space of some time-system is a relation derivative from location in instantaneous spaces of the same time-system. Accordingly location in an instantaneous space is the primary idea which we have to explain. Great confusion has been occasioned in natural philosophy by the neglect to distinguish between the different types of objects, the different types of situation, the different types of location, and the difference between location and situation. It is impossible to reason accurately in the vague concerning objects and their positions without keeping these distinctions in view. An object is located in an abstractive element, when an abstractive set belonging to that element can be found such that each event belonging to that set is a situation of the object. It will be remembered that an abstractive element is a certain group of abstractive sets, and that each abstractive set is a set of events. This definition defines the location of an element in any type of abstractive element. In this sense we can talk of the existence of an object at an instant, meaning thereby its location in some definite moment. It may also be located in some spatial element of the instantaneous space of that moment.

A quantity can be said to be located in an abstractive element when an abstractive set belonging to the element can be found such that the quantitative expressions of the corresponding characters of its events converge to the measure of the given quantity as a limit when we pass along the abstractive set towards its converging end.

By these definitions location in elements of instantaneous spaces is defined. These elements occupy corresponding elements of timeless spaces. An object located in an element of an instantaneous space will also be said to be located at that moment in the timeless element of the timeless space which is occupied by that instantaneous element.

It is not every object which can be located in a moment. An object which can be located in every moment of some duration will be called a 'uniform' object throughout that duration. Ordinary physical objects appear to us to be uniform objects, and we habitually assume that scientific objects such as electrons are uniform. But some sense-objects certainly are not uniform. A tune is an example of a non-uniform object. We have perceived it as a whole in a certain duration; but the tune as a tune is not at any moment of that duration though one of the individual notes may be located there.

It is possible therefore that for the existence of certain sorts of objects, *e.g.* electrons, minimum quanta of time are requisite. Some such postulate is apparently indicated by the modern quantum theory and it is perfectly consistent with the doctrine of objects maintained in these lectures.

Also the instance of the distinction between the electron as the mere quantitative electric charge of its situation and the electron as standing for the ingression of an object throughout nature illustrates the indefinite number of types of objects which exist in nature. We can intellectually distinguish even subtler and subtler types of objects. Here I reckon subtlety as meaning seclusion from the immediate apprehension of sense-awareness. Evolution in the complexity of life means an increase in the types of objects directly sensed. Delicacy of sense-apprehension means perceptions of objects as distinct entities which are mere subtle ideas to cruder sensibilities. The phrasing of music is a mere abstract subtlety to the unmusical; it is a direct sense-apprehension to the initiated. For example, if we could imagine some lowly type of organic being thinking and aware of our thoughts, it would wonder at the abstract subtleties in which we indulge as we think of stones and bricks and drops of water and plants. It only knows of vague undifferentiated feelings in nature. It would consider us as given over to the play of excessively abstract intellects. But then if it could think, it would anticipate; and if it anticipated, it would soon perceive for itself.

In these lectures we have been scrutinising the foundations of natural philosophy. We are stopping at the very point where a boundless ocean of enquiries opens out for our questioning.

I agree that the view of Nature which I have maintained in these lectures is not a simple one. Nature appears as a complex system whose factors are dimly discerned by us. But, as I ask you, Is not this the very truth? Should we not distrust the jaunty assurance with which every age prides itself that it at last has hit upon the ultimate concepts in which all that happens can be formulated? The aim of science is to seek the simplest explanations of complex facts. We are apt to fall into the error of thinking that the facts are simple because simplicity is the goal of our quest. The guiding motto in the life of every natural philosopher should be, Seek simplicity and distrust it.

CHAPTER VIII

SUMMARY

THERE is a general agreement that Einstein's investigations have one fundamental merit irrespective of any criticisms which we may feel inclined to pass on them. They have made us think. But when we have admitted so far, we are most of us faced with a distressing perplexity. What is it that we ought to think about? The purport of my lecture this afternoon will be to meet this difficulty and, so far as I am able, to set in a clear light the changes in the background of our scientific thought which are necessitated by any acceptance, however qualified, of Einstein's main positions. I remember that I am lecturing to the members of a chemical society who are not for the most part versed in advanced mathematics. The first point that I would urge upon you is that what immediately concerns you is not so much the detailed deductions of the new theory as this general change in the background of scientific conceptions which will follow from its acceptance. Of course, the detailed deductions are important, because unless our colleagues the astronomers and the physicists find these predictions to be verified we can neglect the theory altogether. But we may now take it as granted that in many striking particulars these deductions have been found to be in agreement with observation. Accordingly the theory has to be taken seriously and we are anxious to know what will be the consequences of its final acceptance. Furthermore during the last few weeks the scientific journals and the lay press have been filled with articles as to the nature of the crucial experiments which have been made and as to some of the more striking expressions of the outcome of the new theory. 'Space caught bending' appeared on the news-sheet of a well-known evening paper. This rendering is a terse but not inapt translation of Einstein's own way of interpreting his results. I should say at once that I am a heretic as to this explanation and that I shall expound to you another explanation based upon some work of my own, an explanation which seems to me to be more in accordance with our scientific ideas and with the whole body of facts which have to be explained. We have to remember that a new theory must take account of the old well-attested facts of science just as much as of the very latest experimental results which have led to its production.

To put ourselves in the position to assimilate and to criticise any change in ultimate scientific conceptions we must begin at the beginning. So you must bear with me if I commence by making some simple and obvious reflections. Let us consider three statements, (i) 'Yesterday a man was run over on the Chelsea Embankment,' (ii) 'Cleopatra's Needle is on the Charing Cross Embankment,' and (iii) 'There are dark lines in the Solar Spectrum.' The first statement about the accident to the man is about what we may term an 'occurrence,' a 'happening,' or an 'event.' I will use the term 'event' because it is the shortest. In order to specify an observed event, the place, the time, and character of the event are necessary. In specifying the place and the time you are really stating the relation of the assigned event to the general structure of other observed events. For example, the man was run over between your tea and your dinner and adjacently to a passing barge in the river and the traffic in the Strand. The point which I want to make is this: Nature is known to us in our experience as a complex of passing events. In this complex we discern definite mutual relations between component events, which we may call their relative positions, and these positions we express partly in terms of space and partly in terms of time. Also in addition to its mere relative position to other events, each particular event has its own peculiar character. In other words, nature is a structure of events and each event has its position in this structure and its own peculiar character or quality.

Let us now examine the other two statements in the light of this general principle as to the meaning of nature. Take the second statement, 'Cleopatra's Needle is on the Charing Cross Embankment.' At first sight we should hardly call this an event. It seems to lack the element of time or transitoriness. But does it? If an angel had made the remark some hundreds of millions of years ago, the earth was not in existence, twenty millions of years ago there was no Thames, eighty years ago there was no Thames Embankment, and when I was a small boy Cleopatra's Needle was not there. And now that it is there, we none of us expect it to be eternal. The static timeless element in the relation of Cleopatra's Needle to the Embankment is a pure illusion generated by the fact that for purposes of daily intercourse its emphasis is needless. What it comes to is this: Amidst the structure of events which form the medium within which the daily life of Londoners is passed we know how to identify a certain stream of events which maintain permanence of character, namely the character of being the situations of Cleopatra's Needle. Day by day and hour by hour we can find a certain chunk in the transitory life of nature and of that chunk we say, 'There is Cleopatra's Needle.' If we define the Needle in a sufficiently abstract manner we can say that it never changes. But a physicist who looks on that part of the life of nature as a dance of electrons, will tell you that daily it has lost some molecules and gained others, and even the plain man can see that it gets dirtier and is occasionally washed. Thus the question of change in the Needle is a mere matter of definition. The more abstract your definition, the more permanent the Needle. But whether your Needle change or be permanent, all you mean by stating that it is situated on the Charing Cross Embankment, is that amid the structure of events you know of a certain continuous limited stream of events, such that any chunk of that stream, during any hour, or any day, or any second, has the character of being the situation of Cleopatra's Needle.

Finally, we come to the third statement, 'There are dark lines in the Solar Spectrum.' This is a law of nature. But what does that mean? It means merely this. If any event has the character of being an exhibition of the solar spectrum under certain assigned circumstances, it will also have the character of exhibiting dark lines in that spectrum.

This long discussion brings us to the final conclusion that the concrete facts of nature are events exhibiting a certain structure in their mutual relations and certain characters of their own. The aim of science is to express the relations between their characters in terms of the mutual structural relations between the events thus characterised. The mutual structural relations between events are both spatial and temporal. If you think of them as merely spatial you are omitting the temporal element, and if you think of them as merely temporal you are omitting the spatial element. Thus when you think of space alone, or of time alone, you are dealing in abstractions, namely, you are leaving out an essential element in the life of nature as known to you in the experience of your senses. Furthermore there are different ways of making these abstractions which we think of as space and as time; and under some circumstances we adopt one way and under other circumstances we adopt another way. Thus there is no paradox in holding that what we mean by space under one set of circumstances is not what we mean by space under another set of circumstances. And equally what we mean by time under one set of circumstances is not what we mean by time under another set of circumstances. By saying that space and time are abstractions, I do not mean that they do not express for us real facts about nature. What I mean is that there are no spatial facts or temporal facts apart from physical nature, namely that space and time are merely ways of expressing certain truths about the relations between events. Also that under different circumstances there are different sets of truths about the universe which are naturally presented to us as statements about space. In such a case what a being under the one set of circumstances means by space will be different from that meant by a being under the other set of circumstances. Accordingly when we are comparing two observations made under different circumstances we have to ask 'Do the two observers mean the same thing by space and the same thing by time?' The modern theory of relativity has arisen because certain perplexities as to the concordance of certain delicate observations such as the motion of the earth through the ether, the perihelion of mercury, and the positions of the stars in the neighbourhood of the sun, have been solved by reference to this purely relative significance of space and time.

I want now to recall your attention to Cleopatra's Needle, which I have not yet done with. As you are walking along the Embankment you suddenly look up and say, 'Hullo, there's the Needle.' In other words, you recognise it. You cannot recognise an event; because when it is gone, it is gone. You may observe another event of analogous character, but the actual chunk of the life of nature is inseparable from its unique occurrence. But a character of an event can be recognised. We all know that if we go to the Embankment near Charing Cross we shall observe an event having the character which we recognise as Cleopatra's Needle. Things which we thus recognise I call objects. An object is situated in those events or in that stream of events of which it expresses the character. There are many sorts of objects. For example, the colour green is an object according to the above definition. It is the purpose of science to trace the laws which govern the appearance of objects in the various events in which they are found to be situated. For this purpose we can mainly concentrate on two types of objects, which I will call material physical objects and scientific objects. A material physical object is an ordinary bit of matter, Cleopatra's Needle for example. This is a much more complicated type of object than a mere colour, such as the colour of the Needle. I call these simple objects, such as colours or sounds, sense-objects. An artist will train himself to attend more particularly to sense-objects where the ordinary person attends normally to material objects. Thus if you were walking with an artist, when you said 'There's Cleopatra's Needle,' perhaps he simultaneously exclaimed 'There's a nice bit of colour.' Yet you were both expressing your recognition of different component characters of the same event. But in science we have found out that when we know all about the adventures amid events of material physical objects and of scientific objects we have most of the relevant information which will enable us to predict the conditions under which we shall perceive sense-objects in specific situations. For example, when we know that there is a blazing fire (*i.e.* material and scientific objects undergoing various exciting adventures amid events) and opposite to it a mirror (which is another material object) and the positions of a man's face and eyes gazing into the mirror, we know that he can perceive the redness of the flame situated in an event behind the mirror—thus, to a large extent, the appearance of sense-objects is conditioned by the adventures of material objects. The analysis of these adventures makes us aware of another character of events, namely their characters as fields of activity which determine the subsequent events to which they will pass on the objects situated in them. We express these fields of activity in terms of gravitational, electro-magnetic, or chemical forces and attractions. But the exact expression of the nature of these fields of activity forces us intellectually to acknowledge a less obvious type of objects as situated in events. I mean molecules and electrons. These objects are not recognised in isolation. We cannot well miss Cleopatra's Needle, if we are in its neighbourhood; but no one has seen a single molecule or a single electron, yet the characters of events are only explicable to us by expressing them in terms of these scientific objects. Undoubtedly molecules and electrons are abstractions. But then so is Cleopatra's Needle. The concrete facts are the events themselves—I have already explained to you that to be an abstraction does not mean that an entity is nothing. It merely means that its existence is only one factor of a more concrete element of nature. So an electron is abstract because you cannot wipe out the whole structure of events and yet retain the electron in existence. In the same way the grin on the cat is abstract; and the molecule is really in the event in the same sense as the grin is really on the cat's face. Now the more ultimate sciences such as Chemistry or Physics cannot express their ultimate laws in terms of such vague objects as the sun, the earth, Cleopatra's Needle, or a human body. Such objects more properly belong to Astronomy, to Geology, to Engineering, to Archaeology, or to Biology. Chemistry and Physics only deal with them as exhibiting statistical complexes of the effects of their more intimate laws. In a certain sense, they only enter into Physics and Chemistry as technological applications. The reason is that they are too vague. Where does Cleopatra's Needle begin and where does it end? Is the soot part of it? Is it a different object when it sheds a molecule or when its surface enters into chemical combination with the acid of a London fog? The definiteness and permanence of the Needle is nothing to the possible permanent definiteness of a molecule as conceived by science, and the permanent definiteness of a molecule in its turn yields to that of an electron. Thus science in its most ultimate formulation of law seeks objects with the most permanent definite simplicity of character and expresses its final laws in terms of them.

Again when we seek definitely to express the relations of events which arise from their spatio-temporal structure, we approximate to simplicity by progressively diminishing the extent (both temporal and spatial) of the events considered. For example, the event which is the life of the chunk of nature which is the Needle during one minute has to the life of nature within a passing barge during the same minute a very complex spatio-temporal relation. But suppose we progressively diminish the time considered to a second, to a hundredth of a second, to a thousandth of a second, and so on. As we pass along such a series we approximate to an ideal simplicity of structural relations of the pairs of events successively considered, which ideal we call the spatial relations of the Needle to the barge at some instant. Even these relations are too complicated for us, and we consider smaller and smaller bits of the Needle and of the barge. Thus we finally reach the ideal of an event so restricted in its extension as to be without extension in space or extension in time. Such an event is a mere spatial point-flash of instantaneous duration. I call such an ideal event an 'event-particle.' You must not think of the world as ultimately built up of event-particles. That is to put the cart before the horse. The world we know is a continuous stream of occurrence which we can discriminate into finite events forming by their overlappings and containings of each other and separations a spatio-temporal structure. We can express the properties of this structure in terms of the ideal limits to routes of approximation, which I have termed event-particles. Accordingly event-particles are abstractions in their relations to the more concrete events. But then by this time you will have comprehended that you cannot analyse concrete nature without abstracting. Also I repeat, the abstractions of science are entities which are truly in nature, though they have no meaning in isolation from nature.

The character of the spatio-temporal structure of events can be fully expressed in terms of relations between these more abstract event-particles. The advantage of dealing with event-particles is that though they are abstract and complex in respect to the finite events which we directly observe, they are simpler than finite events in respect to their mutual relations. Accordingly they express for us the demands of an ideal accuracy, and of an ideal simplicity in the exposition of relations. These event-particles are the ultimate elements of the four-dimensional space-time manifold which the theory of relativity presupposes. You will have observed that each event-particle is as much an instant of time as it is a point of space. I have called it an instantaneous point-flash. Thus in the structure of this space-time manifold space is not finally discriminated from time, and the possibility remains open for diverse modes of discrimination according to the diverse circumstances of observers. It is this possibility which makes the fundamental distinction between the new way of conceiving the universe and the old way. The secret of understanding relativity is to understand this. It is of no use rushing in with picturesque paradoxes, such as

'Space caught bending,' if you have not mastered this fundamental conception which underlies the whole theory. When I say that it underlies the whole theory, I mean that in my opinion it ought to underlie it, though I may confess some doubts as to how far all expositions of the theory have really understood its implications and its premises.

Our measurements when they are expressed in terms of an ideal accuracy are measurements which express properties of the space-time manifold. Now there are measurements of different sorts. You can measure lengths, or angles, or areas, or volumes, or times. There are also other sorts of measures such as measurements of intensity of illumination, but I will disregard these for the moment and will confine attention to those measurements which particularly interest us as being measurements of space or of time. It is easy to see that four such measurements of the proper characters are necessary to determine the position of an event-particle in the space-time manifold in its relation to the rest of the manifold. For example, in a rectangular field you start from one corner at a given time, you measure a definite distance along one side, you then strike out into the field at right angles, and then measure a definite distance parallel to the other pair of sides, you then rise vertically a definite height and take the time. At the point and at the time which you thus reach there is occurring a definite instantaneous point-flash of nature. In other words, your four measurements have determined a definite event-particle belonging to the four-dimension space-time manifold. These measurements have appeared to be very simple to the land-surveyor and raise in his mind no philosophic difficulties. But

Therapeutic Bodywork

Kathleen Donelan, LMT

For muscle, myofascial, and deep tissue care
pain management and exercise retraining

Exercise Physiologist & LMT
kathatc@gmail.com
www. kathleendonelan.com

Located in Lewes
302-381-4138

suppose there are beings on Mars sufficiently advanced in scientific invention to be able to watch in detail the operations of this survey on earth. Suppose that they construe the operations of the English land-surveyors in reference to the space natural to a being on Mars, namely a Martio-centric space in which that planet is fixed. The earth is moving relatively to Mars and is rotating. To the beings on Mars the operations, construed in this fashion, effect measurements of the greatest complication. Furthermore, according to the relativistic doctrine, the operation of time-measurement on earth will not correspond quite exactly to any time-measurement on Mars.

I have discussed this example in order to make you realise that in thinking of the possibilities of measurement in the space-time manifold, we must not confine ourselves merely to those minor variations which might seem natural to human beings on the earth. Let us make therefore the general statement that four measurements, respectively of independent types (such as measurements of lengths in three directions and a time), can be found such that a definite event-particle is determined by them in its relations to other parts of the manifold.

If (p_1, p_2, p_3, p_4) be a set of measurements of this system, then the event-particle which is thus determined will be said to have p_1, p_2, p_3, p_4 as its co-ordinates in this system of measurement. Suppose that we name it the p-system of measurement. Then in the same p-system by properly varying (p_1, p_2, p_3, p_4) every event-particle that has been, or will be, or instantaneously is now, can be indicated. Furthermore, according to any system of measurement that is natural to us, three of the co-ordinates will be measurements of space and one will be a measurement of time. Let us always take the last co-ordinate to represent the time-measurement. Then we should naturally say that (p_1, p_2, p_3) determined a point in space and that the event-particle happened at that point at the time p_4. But we must not make the mistake of thinking that there is a space in addition to the space-time manifold. That manifold is all that there is for the determination of the meaning of space and time. We have got to determine the meaning of a space-point in terms of the event-particles of the four-dimensional manifold. There is only one way to do this. Note that if we vary the time and take times with the same three space co-ordinates, then the event-particles, thus indicated, are all at the same point. But seeing that there is nothing else except the event-particles, this can only mean that the point (p_1, p_2, p_3) of the space in the p-system is merely the collection of event-particles $(p_1, p_2, p_3, [p_4])$, where p_4 is varied and (p_1, p_2, p_3) is kept fixed. It is rather disconcerting to find that a point in space is not a simple entity; but it is a conclusion which follows immediately from the relative theory of space.

Furthermore the inhabitant of Mars determines event-particles by another system of measurements. Call his system the q-system. According to him (q_1, q_2, q_3, q_4) determines an event-particle, and (q_1, q_2, q_3) determines a point and q_4 a time. But the collection of event-particles which he thinks of as a point is entirely different from any such collection which the man on earth thinks of as a point. Thus the q-space for the man on Mars is quite different from the p-space for the land-surveyor on earth.

So far in speaking of space we have been talking of the timeless space of physical science, namely, of our concept of eternal space in which the world adventures. But the space which we see as we look about is instantaneous space. Thus if our natural perceptions are adjustable to the p-system of measurements we see instantaneously all the event-particles at some definite time p_4, and observe a succession of such spaces as time moves on. The timeless space is achieved by stringing together all these instantaneous spaces. The points of an instantaneous space are event-particles, and the points of an eternal space are strings of event-particles occurring in succession. But the man on Mars will never perceive the same instantaneous spaces as the man on the earth. This system of instantaneous spaces will cut across the earth-man's system. For the earth-man there is one instantaneous space which is the instantaneous present, there are the past spaces and the future spaces. But the present space of the man on Mars cuts across the present space of the man on the earth. So that of the event-particles which the earth-man thinks of as happening now in the present, the man on Mars thinks that some are already past and are ancient history, that others are in the future, and others are in the immediate present. This break-down in the neat conception of a past, a present, and a future is a serious paradox. I call two event-particles which on some or other system of measurement are in the same instantaneous space 'co-present' event-particles. Then it is possible that A and B may be co-present, and that A and C may be co-present, but that B and C may not be co-present. For example, at some inconceivable distance from us there are events co-present with us now and also co-present with the birth of Queen Victoria. If *A* and *B* are co-present there will be some systems in which *A* precedes *B* and some in which *B* precedes *A*. Also there can be no velocity quick enough to carry a material particle from *A* to *B* or from *B* to *A*. These different measure-systems with their divergences of time-reckoning are puzzling, and to some extent affront our common sense. It is not the usual way in which we think of the Universe. We think of one necessary time-system and one necessary space. According to the new theory, there are an indefinite number of discordant time-series and an indefinite number of distinct spaces. Any correlated pair, a time-system and a space-system, will do in which to fit our description of the Universe. We find that under given conditions our measurements are necessarily made in some one pair which together form our natural measure-system. The difficulty as to discordant time-systems is partly solved by distinguishing between what I call the creative advance of nature, which is not properly serial at all, and any one time series. We habitually muddle together this creative advance, which we experience and know as the perpetual transition of nature into novelty, with the single-time series which we naturally employ for measurement. The various time-series each measure some aspect of the creative advance, and the whole bundle of them express all the properties of this advance which are measurable. The reason why we have not previously noted this difference of time-series is the very small difference of properties between any two such series. Any observable phenomena due to this cause depend on the square of the ratio of any velocity entering into the observation to the velocity of light. Now light takes about fifty minutes to get round the earth's orbit; and the earth takes rather more than 17,531 half-hours to do the same. Hence all the effects due to this motion are of the order of the ratio of one to the square of 10,000. Accordingly an earth-man and a sun-man have only neglected effects whose quantitative magnitudes all contain the factor $1/10^8$. Evidently such effects can only be noted by means of the most refined observations. They have been observed however. Suppose we compare two observations on the velocity of light made with the same apparatus as we turn it through a right angle. The velocity of the earth relatively to the sun is in one direction, the velocity of light relatively to the ether should be the same in all directions. Hence if space when we take the ether as at rest means the same thing as space when we take the earth as at rest, we ought to find that the velocity of light relatively to the earth varies according to the direction from which it comes.

These observations on earth constitute the basic principle of the famous experiments designed to detect the motion of the earth through the ether. You all know that, quite unexpectedly, they gave a null result. This is completely explained by the fact that, the space-system and the time-system which we are using are in certain minute ways different from the space and the time relatively to the sun or relatively to any other body with respect to which it is moving.

All this discussion as to the nature of time and space has lifted above our horizon a great difficulty which affects the formulation of all the ultimate laws of physics —for example, the laws of the electromagnetic field, and the law of gravitation. Let us take the law of gravitation as an example. Its formulation is as follows: Two material bodies attract each other with a force proportional to the product of their masses and universely proportional to the square of their distances. In this statement the bodies are supposed to be small enough to be treated as material particles in relation to their distances; and we need not bother further about that minor point. The difficulty to which I want to draw your attention is this: In the formulation of the law one definite time and one definite space are presupposed. The two masses are assumed to be in simultaneous positions.

But what is simultaneous in one time-system may not be simultaneous in another time-system. So according to our new views the law is in this respect not formulated so as to have any exact meaning. Furthermore an analogous difficulty arises over the question of distance. The distance between two instantaneous positions, *i.e.* between two event-particles, is different in different space-systems. What space is to be chosen? Thus again the law lacks precise formulation, if relativity is accepted. Our problem is to seek a fresh interpretation of the law of gravity in which these difficulties are evaded. In the first place we must avoid the abstractions of space and time in the formulation of our fundamental ideas and must recur to the ultimate facts of nature, namely to events. Also in order to find the ideal simplicity of expressions of the relations between events, we restrict ourselves to event-particles. Thus the life of a material particle is its adventure amid a track of event-particles strung out as a continuous series or path in the four-dimensional space-time manifold. These event-particles are the various situations of the material particle. We usually express this fact by adopting our natural space-time system and by talking of the path in space of the material particle as it exists at successive instants of time.

We have to ask ourselves what are the laws of nature which lead the material particle to adopt just this path among event-particles and no other. Think of the path as a whole. What characteristic has that path got which would not be shared by any other slightly varied path? We are asking for more than a law of gravity. We want laws of motion and a general idea of the way to formulate the effects of physical forces.

In order to answer our question we put the idea of the attracting masses in the background and concentrate attention on the field of activity of the events in the neighbourhood of the path. In so doing we are acting in conformity with the whole trend of scientific thought during the last hundred years, which has more and more concentrated attention on the field of force as the immediate agent in directing motion, to the exclusion of the consideration of the immediate mutual influence between two distant bodies. We have got to find the way of expressing the field of activity of events in the neighbourhood of some definite event-particle E of the four-dimensional manifold. I bring in a fundamental physical idea which I call the 'impetus' to express this physical field. The event-particle E is related to any neighbouring event-particle P by an element of impetus. The assemblage of all the elements of impetus relating E to the assemblage of event-particles in the neighbourhood of E expresses the character of the field of activity in the neighbourhood of E. Where I differ from Einstein is that he conceives this quantity which I call the impetus as merely expressing the characters of the space and time to be adopted and thus ends by talking of the gravitational field expressing a curvature in the space-time manifold. I cannot attach any clear conception to his interpretation of space and time. My formulae differ slightly from his, though they agree in those instances where his results have been verified. I need hardly say that in this particular of the formulation of the law of gravitation I have drawn on the general method of procedure which constitutes his great discovery.

Einstein showed how to express the characters of the assemblage of elements of impetus of the field surrounding an event-particle E in terms of ten quantities which I will call J_{11}, J_{12} $(= J_{21})$, J_{22}, J_{23} $(= J_{32})$, etc. It will be noted that there are four spatio-temporal measurements relating E to its neighbour P, and that there are ten pairs of such measurements if we are allowed to take any one measurement twice over to make one such pair. The ten J's depend merely on the position of E in the four-dimensional manifold, and the element of impetus between E and P can be expressed in terms of the ten J's and the ten pairs of the four spatio-temporal measurements relating E and P. The numerical values of the J's will depend on the system of measurement adopted, but are so adjusted to each particular system that the same value is obtained for the element of impetus between E and P, whatever be the system of measurement adopted. This fact is expressed by saying that the ten J's form a 'tensor.' It is not going too far to say that the announcement that physicists would have in future to study the theory of tensors created a veritable panic among them when the verification of Einstein's predictions was first announced.

The ten *J*'s at any event-particle *E* can be expressed in terms of two functions which I call the potential and the 'associate-potential' at *E*. The potential is practically what is meant by the ordinary gravitation potential, when we express ourselves in terms of the Euclidean space in reference to which the attracting mass is at rest. The associate-potential is defined by the modification of substituting the direct distance for the inverse distance in the definition of the potential, and its calculation can easily be made to depend on that of the old-fashioned potential. Thus the calculation of the *J*'s—the coefficients of impetus, as I will call them—does not involve anything very revolutionary in the mathematical knowledge of physicists. We now return to the path of the attracted particle. We add up all the elements of impetus in the whole path, and obtain thereby what I call the 'integral impetus.' The characteristic of the actual path as compared with neighbouring alternative paths is that in the actual paths the integral impetus would neither gain nor lose, if the particle wobbled out of it into a small extremely near alternative path. Mathematicians would express this by saying, that the integral impetus is stationary for an infinitesimal displacement. In this statement of the law of motion I have neglected the existence of other forces. But that would lead me too far afield.

The electromagnetic theory has to be modified to allow for the presence of a gravitational field. Thus Einstein's investigations lead to the first discovery of any relation between gravity and other physical phenomena. In the form in which I have put this modification, we deduce Einstein's fundamental principle, as to the motion of light along its rays, as a first approximation which is absolutely true for infinitely short waves. Einstein's principle, thus partially verified, stated in my language is that a ray of light always follows a path such that the integral impetus along it is zero. This involves that every element of impetus along it is zero.

In conclusion, I must apologise. In the first place I have considerably toned down the various exciting peculiarities of the original theory and have reduced it to a greater conformity with the older physics. I do not allow that physical phenomena are due to oddities of space. Also I have added to the dullness of the lecture by my respect for the audience. You would have enjoyed a more popular lecture with illustrations of delightful paradoxes. But I know also that you are serious students who are here because you really want to know how the new theories may affect your scientific researches.

CHAPTER IX

THE ULTIMATE PHYSICAL CONCEPTS

THE second chapter of this book lays down the first principle to be guarded in framing our physical concept. We must avoid vicious bifurcation Nature is nothing else than the deliverance of sense-awareness. We have no principles whatever to tell us what could stimulate mind towards sense-awareness. Our sole task is to exhibit in one system the characters and inter-relations of all that is observed. Our attitude towards nature is purely 'behaviouristic,' so far as concerns the formulation of physical concepts.

Our knowledge of nature is an experience of activity (or passage). The things previously observed are active entities, the 'events.' They are chunks in the life of nature. These events have to each other relations which in our knowledge differentiate themselves into space-relations and time-relations. But this differentiation between space and time, though inherent in nature, is comparatively superficial; and space and time are each partial expressions of one fundamental relation between events which is neither spatial nor temporal. This relation I call 'extension.' The relation of 'extending over' is the relation of 'including,' either in a spatial or in a temporal sense, or in both. But the mere 'inclusion' is more fundamental than either alternative and does not require any spatio-temporal differentiation. In respect to extension two events are mutually related so that either (i) one includes the other, or (ii) one overlaps the other without complete inclusion, or (iii) they are entirely separate. But great care is required in the definition of spatial and temporal elements from this basis in order to avoid tacit limitations really depending on undefined relations and properties.

Such fallacies can be avoided by taking account of two elements in our experience, namely, (i) our observational 'present,' and (ii) our 'percipient event.'

Our observational 'present' is what I call a 'duration.' It is the whole of nature apprehended in our immediate observation. It has therefore the nature of an event, but possesses a peculiar completeness which marks out such durations as a special type of events inherent in nature. A duration is not instantaneous. It is all that there is of nature with certain temporal limitations. In contradistinction to other events a duration will be called infinite and the other events are finite[1]. In our knowledge of a duration we distinguish (i) certain included events which are particularly discriminated as to their peculiar individualities, and (ii) the remaining included events which are only known as necessarily in being by reason of their relations to the discriminated events and to the whole duration. The duration as a whole is signified[2] by that quality of relatedness (in respect to extension) possessed by the part which is immediately under observation; namely, by the fact that there is essentially a beyond to whatever is observed. I mean by this that every event is known as being related to other events which it does not include. This fact, that every event is known as possessing the quality of exclusion, shows that exclusion is as positive a relation as inclusion. There are of course no merely negative

[1] Cf. note on 'significance,' pp. 197, 198.

[2] Cf. Ch. III, pp. 51 et seq.

relations in nature, and exclusion is not the mere negative of inclusion, though the two relations are contraries. Both relations are concerned solely with events, and exclusion is capable of logical definition in terms of inclusion.

Perhaps the most obvious exhibition of significance is to be found in our knowledge of the geometrical character of events inside an opaque material object. For example we know that an opaque sphere has a centre. This knowledge has nothing to do with the material; the sphere may be a solid uniform billiard ball or a hollow lawn-tennis ball. Such knowledge is essentially the product of significance, since the general character of the external discriminated events has informed us that there are events within the sphere and has also informed us of their geometrical structure.

Some criticisms on 'The Principles of Natural Knowledge' show that difficulty has been found in apprehending durations as real stratifications of nature. I think that this hesitation arises from the unconscious influence of the vicious principle of bifurcation, so deeply embedded in modern philosophical thought. We observe nature as extended in an immediate present which is simultaneous but not instantaneous, and therefore the whole which is immediately discerned or signified as an inter-related system forms a stratification of nature which is a physical fact. This conclusion immediately follows unless we admit bifurcation in the form of the principle of psychic additions, here rejected.

Our 'percipient event' is that event included in our observational present which we distinguish as being in some peculiar way our standpoint for perception. It is roughly speaking that event which is our bodily life within the present duration. The theory of perception as evolved by medical psychology is based on significance. The distant situation of a perceived object is merely known to us as signified by our bodily state, *i.e.* by our percipient event. In fact perception requires sense-awareness of the significations of our percipient event together with sense-awareness of a peculiar relation (situation) between certain objects and the events thus signified. Our percipient event is saved by being the whole of nature by this fact of its significations. This is the meaning of calling the percipient event our standpoint for perception. The course of a ray of light is only derivatively connected with perception. What we do perceive are objects as related to events signified by the bodily states excited by the ray. These signified events (as is the case of images seen behind a mirror) may have very little to do with the actual course of the ray. In the course of evolution those animals have survived whose sense-awareness is concentrated on those significations of their bodily states which are on the average important for their welfare. The whole world of events is signified, but there are some which exact the death penalty for inattention.

The percipient event is always here and now in the associated present duration. It has, what may be called, an absolute position in that duration. Thus one definite duration is associated with a definite percipient event, and we are thus aware of a peculiar relation which finite events can bear to durations. I call this relation 'cogredience.' The notion of rest is derivative from that of cogredience, and the notion of motion is derivative from that of inclusion within a duration without cogredience with it. In fact motion is a relation (of varying character) between an observed event and an observed duration, and cogredience is the most simple character or subspecies of motion. To sum up, a duration and a percipient event are essentially involved in the general character of each observation of nature, and the percipient event is cogredient with the duration.

Our knowledge of the peculiar characters of different events depends upon our power of comparison. I call the exercise of this factor in our knowledge 'recognition,' and the requisite sense-awareness of the comparable characters I call 'sense-recognition.' Recognition and abstraction essentially involve each other. Each of them exhibits an entity for knowledge which is less than the concrete fact, but is a real factor in that fact. The most concrete fact capable of separate discrimination is the event. We cannot abstract without recognition, and we cannot recognise without abstraction. Perception involves apprehension of the event and recognition of the factors of its character.

The things recognised are what I call 'objects.' In this general sense of the term the relation of extension is itself an object. In practice however I restrict the term to those objects which can in some sense or other be said to have a situation in an event; namely, in the phrase 'There it is again' I restrict the 'there' to be the indication of a special event which is the situation of the object. Even so, there are different types of objects, and statements which are true of objects of one type are not in general true of objects of other types. The objects with which we are here concerned in the formulation of physical laws are material objects, such as bits of matter, molecules and electrons. An object of one of these types has relations to events other than those belonging to the stream of its situations. The fact of its situations within this stream has impressed on all other events certain modifications of their characters. In truth the object in its completeness may be conceived as a specific set of correlated modifications of the characters of all events, with the property that these modifications attain to a certain focal property for those events which belong to the stream of its situations. The total assemblage of the modifications of the characters of events due to the existence of an object in a stream of situations is what I call the 'physical field' due to the object. But the object cannot really be separated from its field. The object is in fact nothing else than the systematically adjusted set of modifications of the field. The conventional limitation of the object to the focal stream of events in which it is said to be 'situated' is convenient for some purposes, but it obscures the ultimate fact of nature. From this point of view the antithesis between action at a distance and action by transmission is meaningless. The doctrine of this paragraph is nothing else than another way of expressing the unresolvable multiple relation of an object to events.

A complete time-system is formed by any one family of parallel durations. Two durations are parallel if either (i) one includes the other, or (ii) they overlap so as to include a third duration common to both, or (iii) are entirely separate. The excluded case is that of two durations overlapping so as to include in common an aggregate of finite events but including in common no other complete duration. The recognition of the fact of an indefinite number of families of parallel durations is what differentiates the concept of nature here put forward from the older orthodox concept of the essentially unique time-systems. Its divergence from Einstein's concept of nature will be briefly indicated later.

The instantaneous spaces of a given time-system are the ideal (non-existent) durations of zero temporal thickness indicated by routes of approximation along series formed by durations of the associated family. Each such instantaneous space represents the ideal of nature at an instant and is also a moment of time. Each time-system thus possesses an aggregate of moments belonging to it alone. Each event-particle lies in one and only one moment of a given time-system. An event-particle has three characters[1]: (i) its extrinsic character which is its character as a definite route of convergence among events, (ii) its intrinsic character which is the peculiar quality of nature in its neighbourhood, namely, the character of the physical field in the neighbourhood, and (iii) its position.

The position of an event-particle arises from the aggregate of moments (no two of the same family) in which it lies. We fix our attention on one of these moments which is approximated to by the short duration of our immediate experience, and we express position as the position in this moment. But the event-particle receives its position in moment M in virtue of the whole aggregate of other moments M', M'', etc., in which it also lies. The differentiation of M into a geometry of event-particles (instantaneous points) expresses the differentiation of M by its intersections with moments of alien time-systems. In this way planes and straight lines and event-particles themselves find their being. Also the parallelism of planes and straight lines arises from the parallelism of the moments of one and

[1] Cf. pp. 82 et seq.

the same time-system intersecting M. Similarly the order of parallel planes and of event-particles on straight lines arises from the time-order of these intersecting moments. The explanation is not given here[1]. It is sufficient now merely to mention the sources from which the whole of geometry receives its physical explanation.

The correlation of the various momentary spaces of one time-system is achieved by the relation of cogredience. Evidently motion in an instantaneous space is unmeaning. Motion expresses a comparison between position in one instantaneous space with positions in other instantaneous spaces of the same time-system. Cogredience yields the simplest outcome of such comparison, namely, rest.

Motion and rest are immediately observed facts. They are relative in the sense that they depend on the time-system which is fundamental for the observation. A string of event-particles whose successive occupation means rest in the given time-system forms a timeless point in the timeless space of that time-system. In this way each time-system possesses its own permanent timeless space peculiar to it alone, and each such space is composed of timeless points which belong to that time-system and to no other. The paradoxes of relativity arise from neglecting the fact that different assumptions as to rest involve the expression of the facts of physical science in terms of radically different spaces and times, in which points and moments have different meanings.

The source of order has already been indicated and that of congruence is now found. It depends on motion.

[1] Cf. *Principles of Natural Knowledge*, and previous chapters of the present work.

From cogredience, perpendicularity arises; and from perpendicularity in conjunction with the reciprocal symmetry between the relations of any two time-systems congruence both in time and space is completely defined (cf. *loc. cit.*).

The resulting formulae are those for the electromagnetic theory of relativity, or, as it is now termed, the restricted theory. But there is this vital difference: the critical velocity c which occurs in these formulae has now no connexion whatever with light or with any other fact of the physical field (in distinction from the extensional structure of events). It simply marks the fact that our congruence determination embraces both times and spaces in one universal system, and therefore if two arbitrary units are chosen, one for all spaces and one for all times, their ratio will be a velocity which is a fundamental property of nature expressing the fact that times and spaces are really comparable.

The physical properties of nature are expressed in terms of material objects (electrons, etc.). The physical character of an event arises from the fact that it belongs to the field of the whole complex of such objects. From another point of view we can say that these objects are nothing else than our way of expressing the mutual correlation of the physical characters of events.

The spatio-temporal measurableness of nature arises from (i) the relation of extension between events, and (ii) the stratified character of nature arising from each of the alternative time-systems, and (iii) rest and motion, as exhibited in the relations of finite events to time-systems. None of these sources of measurement depend on the physical characters of finite events as exhibited by the situated objects. They are completely signified for events whose physical characters are unknown. Thus the spatio-temporal measurements are independent of the objectival physical characters. Furthermore the character of our knowledge of a whole duration, which is essentially derived from the significance of the part within the immediate field of discrimination, constructs it for us as a uniform whole independent, so far as its extension is concerned, of the unobserved characters of remote events. Namely, there is a definite whole of nature, simultaneously now present, whatever may be the character of its remote events. This consideration reinforces the previous conclusion. This conclusion leads to the assertion of the essential uniformity of the momentary spaces of the various time-systems, and thence to the uniformity of the timeless spaces of which there is one to each time-system.

The analysis of the general character of observed nature set forth above affords explanations of various fundamental observational facts: (α) It explains the differentiation of the one quality of extension into time and space. (β) It gives a meaning to the observed facts of geometrical and temporal position, of geometrical and temporal order, and of geometrical straightness and planeness. (γ) It selects one definite system of congruence embracing both space and time, and thus explains the concordance as to measurement which is in practice attained. (δ) It explains (consistently with the theory of relativity) the observed phenomena of rotation, *e.g.* Foucault's pendulum, the equatorial bulge of the earth, the fixed senses of rotation of cyclones and anticyclones, and the gyro-compass. It does this by its admission of definite stratifications of nature which are disclosed by the very character of our knowledge of it. (ϵ) Its explanations of motion are more fundamental than those expressed in (δ); for it explains what is meant by motion itself. The observed motion of an extended object is the relation of its various situations to the stratification of nature expressed by the time-system fundamental to the observation. This motion expresses a real relation of the object to the rest of nature. The quantitative expression of this relation will vary according to the time-system selected for its expression.

This theory accords no peculiar character to light beyond that accorded to other physical phenomena such as sound. There is no ground for such a differentiation. Some objects we know by sight only, and other objects we know by sound only, and other objects we observe neither by light nor by sound but by touch or smell or otherwise. The velocity of light varies according to its medium and so does that of sound. Light moves in curved paths under certain conditions and so does sound. Both light and sound are waves of disturbance in the physical characters of events; and (as has been stated above, p. 188) the actual course of the light is of no more importance for perception than is the actual course of the sound. To base the whole philosophy of nature upon light is a baseless assumption. The Michelson-Morley and analogous experiments show that within the limits of our inexactitude of observation the velocity of light is an approximation to the critical velocity 'c' which expresses the relation between our space and time units. It is provable that the assumption as to light by which these experiments and the influence of the gravitational field on the light-rays are explained is deducible *as an approximation* from the equations of the electromagnetic field. This completely disposes of any necessity for differentiating light from other physical phenomena as possessing any peculiar fundamental character.

It is to be observed that the measurement of extended nature by means of extended objects is meaningless apart from some observed fact of simultaneity inherent in nature and not merely a play of thought. Otherwise there is no meaning to the concept of one presentation of your extended measuring rod AB. Why not AB' where B' is the end B five minutes later? Measurement presupposes for its possibility nature as a simultaneity, and an observed object present then and present now. In other words, measurement of extended nature requires some inherent character in nature affording a rule of presentation of events. Furthermore congruence cannot be defined by the permanence of the measuring rod. The permanence is itself meaningless apart from some immediate judgment of self-congruence. Otherwise how is an elastic string differentiated from a rigid measuring rod? Each remains the same self-identical object. Why is one a possible measuring rod and the other not so? The meaning of congruence lies beyond the self-identity of the object. In other words measurement presupposes the measurable, and the theory of the measurable is the theory of congruence.

Furthermore the admission of stratifications of nature bears on the formulation of the laws of nature. It has been laid down that these laws are to be expressed in differential equations which, as expressed in any general system of measurement, should bear no reference to any other particular measure-system. This requirement is purely arbitrary. For a measure-system measures something inherent in nature; otherwise it has no connexion with nature at all. And that something which is measured by a particular measure-system may have a special relation to the phenomenon whose law is being formulated. For example the gravitational field due to a material object at rest in a certain time-system may be expected to exhibit in its formulation particular reference to spatial and temporal quantities of that time-system. The field can of course be expressed in any measure-systems, but the particular reference will remain as the simple physical explanation.

NOTE: ON THE GREEK CONCEPT OF A POINT

The preceding pages had been passed for press before I had the pleasure of seeing Sir T. L. Heath's *Euclid in Greek*[1]. In the original Euclid's first definition is

σημεῖόν ἐστιν, οὗ μέρος οὐθέν.

I have quoted it on p. 86 in the expanded form taught to me in childhood, 'without parts and without magnitude.' I should have consulted Heath's English edition—a classic from the moment of its issue—before committing myself to a statement about Euclid. This is however a trivial correction not affecting sense and not worth a note. I wish here to draw attention to Heath's own note to this definition in his *Euclid in Greek*. He summarises Greek thought on the nature of a point, from the Pythagoreans, through Plato and Aristotle, to Euclid. My analysis of the requisite character of a point on pp. 89 and 90 is in complete agreement with the outcome of the Greek discussion.

NOTE: ON SIGNIFICANCE AND INFINITE EVENTS

The theory of significance has been expanded and made more definite in the present volume. It had already been introduced in the *Principles of Natural Knowledge* (cf. subarticles 3·3 to 3·8 and 16·1, 16·2, 19·4, and articles 20, 21). In reading over the proofs of the present volume, I come to the conclusion that in the

[1] Camb. Univ. Press, 1920.

light of this development my limitation of infinite events to durations is untenable. This limitation is stated in article 33 of the *Principles* and at the beginning of Chapter IV (p. 74) of this book. There is not only a significance of the discerned events embracing the whole present duration, but there is a significance of a cogredient event involving its extension through a whole time-system backwards and forwards. In other words the essential 'beyond' in nature is a definite beyond in time as well as in space [cf. pp. 53, 194]. This follows from my whole thesis as to the assimilation of time and space and their origin in extension. It also has the same basis in the analysis of the character of our knowledge of nature. It follows from this admission that it is possible to define point-tracks [*i.e.* the points of timeless spaces] as abstractive elements. This is a great improvement as restoring the balance between moments and points. I still hold however to the statement in subarticle 35·4 of the *Principles* that the intersection of a pair of non-parallel durations does not present itself to us as one event. This correction does not affect any of the subsequent reasoning in the two books.

I may take this opportunity of pointing out that the 'stationary events' of article 57 of the *Principles* are merely cogredient events got at from an abstract mathematical point of view.

INDEX

In the case of terms of frequent occurrence, only those occurrences are indexed which are of peculiar importance for the elucidation of meaning.

COSIMO is a specialty publisher of books and publications that inspire, inform and engage readers. Our mission is to offer unique books to niche audiences around the world.

COSIMO CLASSICS offers a collection of distinctive titles by the great authors and thinkers throughout the ages. At **COSIMO CLASSICS** timeless classics find a new life as affordable books, covering a variety of subjects including: *Biographies, Business, History, Mythology, Personal Development, Philosophy, Religion and Spirituality,* and much more!

COSIMO-on-DEMAND publishes books and publications for innovative authors, non-profit organizations and businesses. **COSIMO-on-DEMAND** specializes in bringing books back into print, publishing new books quickly and effectively, and making these publications available to readers around the world.

COSIMO REPORTS publishes public reports that affect your world: from global trends to the economy, and from health to geo-politics.

FOR MORE INFORMATION CONTACT US AT
INFO@COSIMOBOOKS.COM

If you are a book-lover interested in our current catalog of books.

If you are an author who wants to get published

If you represent an organization or business seeking to reach your members, donors or customers with your own books and publications

COSIMO BOOKS ARE ALWAYS AVAILABLE AT ONLINE BOOKSTORES

VISIT COSIMOBOOKS.COM
BE INSPIRED, BE INFORMED

Printed in the United States
206994BV00001B/197/A

9 781602 062139